Comparing Groups

Comparing Groups

Randomization and Bootstrap Methods Using R

Andrew S. Zieffler
University of Minnesota

Jeffrey R. Harring
University of Maryland

Jeffrey D. Long
University of Iowa

A JOHN WILEY & SONS, INC., PUBLICATION

Library of Congress Cataloging-in-Publication Data:

Zieffler, Andrew, 1974–
  Comparing groups : randomization and bootstrap methods using R / Andrew Zieffler, Jeffrey Harring, Jeffrey D. Long.
     p. cm.
   Includes bibliographical references and index.
   ISBN 978-0-470-62169-1 (hardback)
1. Bootstrap (Statistics) 2. Random data (Statistics) 3. Psychology—Data processing. 4. R (Computer program language) 5. Distribution (Probability theory) I. Harring, Jeffrey, 1964– II. Long, Jeffrey D., 1964– III. Title.
   QA276.8.Z53 2011
   519.5'4—dc22   2010054064

10 9 8 7 6 5 4 3 2 1

CONTENTS

List of Figures xiii

List of Tables xxi

Foreword xxiii

Preface xxv

Acknowledgments xxxi

1 An Introduction to R **1**

 1.1 Getting Started 2

   1.1.1 Windows OS 2

   1.1.2 Mac OS 2

   1.1.3 Add-On Packages 2

 1.2 Arithmetic: R as a Calculator 4

 1.3 Computations in R: Functions 4

 1.4 Connecting Computations 7

   1.4.1 Naming Conventions 8

 1.5 Data Structures: Vectors 9

   1.5.1 Creating Vectors in R 9

   1.5.2 Computation with Vectors 11

   1.5.3 Character and Logical Vectors 12

1.6	Getting Help	13
1.7	Alternative Ways to Run R	14
1.8	Extension: Matrices and Matrix Operations	14
	1.8.1 Computation with Matrices	15
1.9	Further Reading	18
	Problems	19

2   Data Representation and Preparation   **21**

2.1	Tabular Data	23
	2.1.1 External Formats for Storing Tabular Data	23
2.2	Data Entry	24
	2.2.1 Data Codebooks	25
2.3	Reading Delimited Data into R	25
	2.3.1 Identifying the Location of a File	26
	2.3.2 Examining the Data in a Text Editor	28
	2.3.3 Reading Delimited Separated Data: An Example	28
2.4	Data Structure: Data Frames	29
	2.4.1 Examining the Data Read into R	29
2.5	Recording Syntax using Script Files	33
	2.5.1 Documentation File	34
2.6	Simple Graphing in R	34
	2.6.1 Saving Graphics to Insert into a Word-Processing File	35
2.7	Extension: Logical Expressions and Graphs for Categorical Variables	37
	2.7.1 Logical Operators	38
	2.7.2 Measurement Level and Analysis	40
	2.7.3 Categorical Data	42
	2.7.4 Plotting Categorical Data	44
2.8	Further Reading	45
	Problems	46

3   Data Exploration: One Variable   **49**

3.1	Reading In the Data	50
3.2	Nonparametric Density Estimation	52
	3.2.1 Graphically Summarizing the Distribution	52
	3.2.2 Histograms	52
	3.2.3 Kernel Density Estimators	53
	3.2.4 Controlling the Density Estimation	53

	3.2.5	Plotting the Estimated Density	55
3.3		Summarizing the Findings	58
	3.3.1	Creating a Plot for Publication	59
	3.3.2	Writing Up the Results for Publication	61
3.4		Extension: Variability Bands for Kernel Densities	62
3.5		Further Reading	62
		Problems	63

4 Exploration of Multivariate Data: Comparing Two Groups 65

4.1		Graphically Summarizing the Marginal Distribution	66
4.2		Graphically Summarizing Conditional Distributions	66
	4.2.1	Indexing: Accessing Individuals or Subsets	68
	4.2.2	Indexing Using a Logical Expression	69
	4.2.3	Density Plots of the Conditional Distributions	70
	4.2.4	Side-by-Side Box-and-Whiskers Plots	70
4.3		Numerical Summaries of Data: Estimates of the Population Parameters	72
	4.3.1	Measuring Central Tendency	73
	4.3.2	Measuring Variation	74
	4.3.3	Measuring Skewness	76
	4.3.4	Kurtosis	78
4.4		Summarizing the Findings	80
	4.4.1	Creating a Plot for Publication	80
	4.4.2	Using Color	81
	4.4.3	Selecting a Color Palette	85
4.5		Extension: Robust Estimation	87
	4.5.1	Robust Estimate of Location: The Trimmed Mean	87
	4.5.2	Robust Estimate of Variation: The Winsorized Variance	89
4.6		Further Reading	91
		Problems	91

5 Exploration of Multivariate Data: Comparing Many Groups 93

5.1		Graphing Many Conditional Distributions	94
	5.1.1	Panel Plots	96
	5.1.2	Side-by-Side Box-and-Whiskers Plots	97
5.2		Numerically Summarizing the Data	100
5.3		Summarizing the Findings	101
	5.3.1	Writing Up the Results for Publication	102

		5.3.2	Enhancing a Plot with a Line	102
5.4			Examining Distributions Conditional on Multiple Variables	103
5.5			Extension: Conditioning on Continuous Variables	107
		5.5.1	Scatterplots of the Conditional Distributions	110
5.6			Further Reading	112
			Problems	113

6   Randomization and Permutation Tests   115

6.1			Randomized Experimental Research	118
6.2			Introduction to the Randomization Test	119
6.3			Randomization Tests with Large Samples: Monte Carlo Simulation	122
		6.3.1	Rerandomization of the Data	124
		6.3.2	Repeating the Randomization Process	125
		6.3.3	Generalizing Processes: Functions	126
		6.3.4	Repeated Operations on Matrix Rows or Columns	127
		6.3.5	Examining the Monte Carlo Distribution and Obtaining the p-Value	127
6.4			Validity of the Inferences and Conclusions Drawn from a Randomization Test	130
		6.4.1	Exchangeability	130
		6.4.2	Nonexperimental Research: Permutation Tests	131
		6.4.3	Nonexperimental, Nongeneralizable Research	131
6.5			Generalization from the Randomization Results	132
6.6			Summarizing the Results for Publication	133
6.7			Extension: Tests of the Variance	133
6.8			Further Reading	134
			Problems	135

7   Bootstrap Tests   137

7.1			Educational Achievement of Latino Immigrants	138
7.2			Probability Models: An Interlude	140
7.3			Theoretical Probability Models in R	141
7.4			Parametric Bootstrap Tests	143
		7.4.1	Choosing a Probability Model	144
		7.4.2	Standardizing the Distribution of Achievement Scores	144
7.5			The Parametric Bootstrap	146

	7.5.1	The Parametric Bootstrap: Approximating the Distribution of the Mean Difference	146
7.6		Implementing the Parametric Bootstrap in R	148
	7.6.1	Writing a Function to Randomly Generate Data for the boot() Function	148
	7.6.2	Writing a Function to Compute a Test Statistic Using the Randomly Generated Data	150
	7.6.3	The Bootstrap Distribution of the Mean Difference	151
7.7		Summarizing the Results of the Parametric Bootstrap Test	154
7.8		Nonparametric Bootstrap Tests	154
	7.8.1	Using the Nonparametric Bootstrap to Approximate the Distribution of the Mean Difference	157
	7.8.2	Implementing the Nonparametric Bootstrap in R	158
7.9		Summarizing the Results for the Nonparametric Bootstrap Test	160
7.10		Bootstrapping Using a Pivot Statistic	161
	7.10.1	Student's t-Statistic	161
7.11		Independence Assumption for the Bootstrap Methods	164
7.12		Extension: Testing Functions	166
	7.12.1	Ordering a Data Frame	166
7.13		Further Reading	168
		Problems	168

8   Philosophical Considerations   171

8.1		The Randomization Test vs. the Bootstrap Test	172
8.2		Philosophical Frameworks of Classical Inference	173
	8.2.1	Fisher's Significance Testing	174
	8.2.2	Neyman–Pearson Hypothesis Testing	175
	8.2.3	p-Values	176

9   Bootstrap Intervals and Effect Sizes   179

9.1		Educational Achievement Among Latino Immigrants: Example Revisited	180
9.2		Plausible Models to Reproduce the Observed Result	180
	9.2.1	Computing the Likelihood of Reproducing the Observed Result	181
9.3		Bootstrapping Using an Alternative Model	185
	9.3.1	Using R to Bootstrap under the Alternative Model	187

9.3.2   Using the Bootstrap Distribution to Compute the Interval Limits | 190

9.3.3   Historical Interlude: Student's Approximation for the Interval Estimate | 190

9.3.4   Studentized Bootstrap Interval | 191

9.4   Interpretation of the Interval Estimate | 191

9.5   Adjusted Bootstrap Intervals | 192

9.6   Standardized Effect Size: Quantifying the Group Differences in a Common Metric | 192

9.6.1   Effect Size as Distance—Cohen's δ | 193

9.6.2   Robust Distance Measure of Effect | 195

9.7   Summarizing the Results | 197

9.8   Extension: Bootstrapping the Confidence Envelope for a Q-Q Plot | 197

9.9   Confidence Envelopes | 198

9.10   Further Reading | 202

Problems | 204

10   Dependent Samples

10   Dependent Samples | **205**

10.1   Matching: Reducing the Likelihood of Nonequivalent Groups | 206

10.2   Mathematics Achievement Study Design | 206

10.2.1   Exploratory Analysis | 209

10.3   Randomization/Permutation Test for Dependent Samples | 211

10.3.1   Reshaping the Data | 212

10.3.2   Randomization Test Using the Reshaped Data | 214

10.4   Effect Size | 216

10.5   Summarizing the Results of a Dependent Samples Test for Publication | 217

10.6   To Match or Not to Match . . . That is the Question | 218

10.7   Extension: Block Bootstrap | 220

10.8   Further Reading | 223

Problems | 224

11   Planned Contrasts | **227**

11.1   Planned Comparisons | 228

11.2   Examination of Weight Loss Conditioned on Diet | 228

11.2.1   Exploration of Research Question 1 | 229

11.2.2   Exploration of Research Question 2 | 230

11.2.3   Exploration of Research Question 3   231

11.3   From Research Questions to Hypotheses   232

11.4   Statistical Contrasts   233

11.4.1   Complex Contrasts   236

11.5   Computing the Estimated Contrasts Using the Observed Data   237

11.6   Testing Contrasts: Randomization Test   239

11.7   Strength of Association: A Measure of Effect   240

11.7.1   Total Sum of Squares   241

11.8   Contrast Sum of Squares   243

11.9   Eta-Squared for Contrasts   243

11.10 Bootstrap Interval for Eta-Squared   244

11.11 Summarizing the Results of a Planned Contrast Test Analysis   245

11.12 Extension: Orthogonal Contrasts   245

11.13 Further Reading   251

Problems   251

12   Unplanned Contrasts   253

12.1   Unplanned Comparisons   254

12.2   Examination of Weight Loss Conditioned on Diet   254

12.3   Omnibus Test   257

12.3.1   Statistical Models   257

12.3.2   Postulating a Statistical Model to Fit the Data   258

12.3.3   Fitting a Statistical Model to the Data   260

12.3.4   Partitioning Variation in the Observed Scores   262

12.3.5   Randomization Test for the Omnibus Hypothesis   268

12.4   Group Comparisons After the Omnibus Test   269

12.5   Ensemble-Adjusted p-values   270

12.5.1   False Discovery Rate   272

12.6   Strengths and Limitations of the Four Approaches   273

12.6.1   Planned Comparisons   273

12.6.2   Omnibus Test Followed by Unadjusted Group
Comparisons   274

12.6.3   Omnibus Test Followed by Adjusted Group
Comparisons   274

12.6.4   Adjusted Group Comparisons without the Omnibus Test   275

12.6.5   Final Thoughts   276

12.7   Summarizing the Results of Unplanned Contrast Tests for
Publication   276

12.8	Extension: Plots of the Unplanned Contrasts	276
	12.8.1 Simultaneous Intervals	280
12.9	Further Reading	282
	Problems	283

References   **285**

LIST OF FIGURES

1.1   Example of the R console window on a Mac.   3

2.1   MRI scan stored as an image file.   22

2.2   Latino education data opened in *Text Editor* on the Mac. The data are delimited by commas and each case is delimited by a line break   24

2.3   Folder and file hierarchy on a computer system.   27

2.4   Box-and-whiskers plot of the distribution of educational achievement scores for 150 Latino immigrants.   36

2.5   Density plot of the distribution of educational achievement scores for 150 Latino immigrants.   36

2.6   Density plot of the distribution of educational achievement scores for 41 Latinos who immigrated to the United States between the ages of 4 and 7.   41

2.7   Bar plot showing the frequencies for Latinos in the sample who are fluent and not fluent in English.   45

2.8   Bar plot showing the proportions for Latinos in the sample who are fluent and not fluent in English.   45

3.1   Illustration of the kernel density estimation (solid line) for $N = 6$ observations (vertical lines). A Gaussian kernel function (dashed lines) with a fixed smoothing parameter was centered at each observation. The figure was adapted from Sain (1994).   54

3.2   Kernel density estimates using three different smoothing parameters. Each estimate used a Gaussian kernel. The estimate on the left used a smoothing parameter of 0.5. The estimate in the middle used a smoothing parameter of 2. The estimate on the right used a smoothing parameter of 10.   55

3.3   Kernel density estimates using Silverman's rule of thumb for the smoothing parameter (left-hand plot). This is the default value for the smoothing parameter. The middle plot uses the Sheather and Jones "solve the equation" method of computing the smoothing parameter, and the right-hand plot uses an adaptive kernel function.   57

3.4   Kernel density estimate for the distribution of ages for a sample of 28,633 Vietnamese citizens using an adaptive kernel method.   59

3.5   Kernel density estimate (solid line) with variability bands (dotted lines) for the distribution of ages for the Vietnamese population. A smoothing parameter of $h = 4$ was used in the density estimation.   63

4.1   Kernel density estimate for the marginal distribution of household per capita expenditures (in U.S. dollars).   68

4.2   Kernel density plots for the distribution of household per capita expenditures (in dollars) conditioned on area. The density of household per capita expenditures for urban households is plotted using a solid line and that for rural households is plotted using a dotted line.   71

4.3   Side-by-side box-and-whiskers plots for the conditional distributions of household per capita expenditures (in U.S. dollars) using the `names=` argument.   72

4.4   Simulated density plots for the distribution of household per capita expenditures (in dollars) conditioned on area showing large within-group variation.   75

4.5   Simulated density plots for the distribution of household per capita expenditures (in dollars) conditioned on area showing small within-group variation.   75

4.6   Kernel density estimate for a mesokurtic distribution (solid line) and a leptokurtic distribution (dotted line). The leptokurtic distributions are skinnier and more peaked than the mesokurtic distribution.   79

4.7   Kernel density estimate for a mesokurtic distribution (solid line) and a platykurtic distribution (dotted line). The platykurtic distribution is flatter than the mesokurtic distribution.   79

4.8   Plots of the kernel density estimates for the distribution of household per capita expenditures (in dollars) for the rural ($n = 4269$) and urban households ($n = 1730$).   82

4.9   Kernel density plots for the distribution of household per capita expenditures (in dollars) conditioned on area. The density of household per capita expenditures for both distributions are shaded using the polygon() function. The plot appears in grayscale in this book, but should print in color on a computer.   84

5.1   Kernel density plots for the distribution of household per capita expenditures (in dollars) conditioned on region. The graph is problematic as the various regions are difficult to sort out. The plot appears in grayscale in this book, but should print in color on a computer.   96

5.2   The default panel array is mfrow=c(1,1). This array has one row and one column.   97

5.3   This panel array is mfrow=c(2,3). This array has two rows and three columns.   97

5.4   Panel plot of the density of household per capita expenditures (in dollars) conditioned on region.   99

5.5   Side-by-side box-and-whiskers plots of household per capita expenditures (in U.S. dollars) conditioned on region.   100

5.6   Plot of the estimated density of household per capita expenditures (in U.S. dollars) conditioned on region with the poverty line (shown as a dashed line) demarcated.   104

5.7   Plotted density estimates of the per capita household expenditures for the urban households (solid line) and rural households (dotted line) conditioned on region. The poverty line of $119 is expressed as a vertical dashed line in each panel.   106

5.8   The distribution of mathematics achievement conditioned on average weekly time spent on mathematics homework. Achievement is reported as a T-score which are transformed scores having a mean of 50 and a standard deviation of 10.   110

5.9   Scatterplot showing the distribution of mathematics achievement conditioned on average weekly time spent on mathematics homework.   112

5.10   Scatterplot and side-by-side box-and-whiskers plots showing the distribution of mathematics achievement conditioned on average weekly time spent on mathematics homework.   112

6.1   Kernel density plots for the distribution of the T-scaled delinquency measure for students who participated in the after-school program (solid line) and for students who did not (dashed line).   124

6.2   Kernel density plot for the distribution of permuted mean differences. The point represents the observed mean difference of 1.7. The shaded area represents the p-value.   129

7.1   Density plots for the distribution of educational achievement conditioned on whether or not the Latin American country of emmigration was Mexico.   140

7.2   Side-by-side box-and-whiskers plots for the distribution of educational achievement conditioned on whether or not the Latin American country of emmigration was Mexico.   140

7.3   Spike plot of the probability distribution of the outcome for the number of heads in 10 tossed coins.   143

7.4   Density plots for the distribution of standardized educational achievement conditioned on whether or not the Latin American country of emmigration was Mexico.   147

7.5   Side-by-side box-and-whiskers plots for the distribution of standardized educational achievement conditioned on whether or not the Latin American country of emmigration was Mexico.   147

7.6   Steps for bootstrapping the distribution for a particular test statistic, $\hat{V}$, from a known probability distribution.   148

7.7   Visualization of the process used to parametrically bootstrap 150 observations from the standard normal distribution. The first 34 observations, the X replicates, constitute the replicated "non-Mexican immigrants" and the last 116 observations, the Y replicates, constitute the replicated "Mexican immigrants." $\hat{V}$ is the test statistic (e.g., the mean difference) computed for each replicate data set.   149

7.8   Bootstrap distribution of the mean difference in achievement scores between non-Mexican and Mexican immigrants for 4999 replicates under the hypothesis of no difference. The vertical line is drawn at 0.   153

7.9   Density plot (dashed line) with variability bands of the distribution of standardized educational achievement for Mexican immigrants. The solid line is the density curve of the standard normal probability model.   155

7.10   Density plot (dashed line) with variability bands of the distribution of standardized educational achievement for non-Mexican immigrants. The solid line is the density curve of the standard normal probability model.   155

7.11   Nonparametric bootstrapping involves resampling from the pooled observed sample with replacement. The bootstrap distribution of a statistic obtained from this method is typically a good approximation of the exact distribution.   156

7.12   Steps for using the sample data to bootstrap the distribution for a particular test statistic, $\hat{V}$, when the probability distribution is unknown.   157

7.13   Bootstrap distribution of the mean difference (using nonparametric bootstrapping) in achievement scores between non-Mexican and Mexican immigrants for 4999 replicates assuming there are no population differences. The vertical line is drawn at zero.   159

9.1   The bootstrap distributions of the standardized mean difference based on the model with parameter value 0 (dotted), parameter value 3 (dashed), and parameter value 10 (dot-dashed). The point in the plot demarcates the observed mean difference in the data of 5.9.   182

9.2   The bootstrap distribution of the mean difference based on the model with parameter value 3. A vertical dashed line is drawn at 3, the model's parameter value. The point in the plot demarcates the observed mean difference in the data of 5.9. The one-sided *p*-value, indicating the likelihood of this model to reproduce the observed data, is shaded.   184

9.3   Nonparametric bootstrapping under an alternative model involves resampling from the observed sample with replacement within defined strata.   187

9.4   Steps for using the sample data to bootstrap the distribution for a particular test statistic, $\hat{V}$, under an alternative model.   188

9.5   Bootstrap distribution of the mean difference in educational achievement scores between non-Mexican and Mexican immigrants for 4999 replicates. The resampling was carried out under the alternative model, thus centering the distribution around the observed mean difference of 5.9 (dashed line).   190

9.6   Density plot of the 4999 bootstrapped robust standardized effect sizes. The resampling was carried out under the alternative model, thus centering the distribution around the observed robust standardized effect of 0.38 (vertical dashed line).   198

9.7   Q-Q plot of the standardized Mexican achievement scores. The line references where the points would lie if the sample of scores were normally distributed.   199

9.8   Density plot of the empirical standardized Mexican achievement scores. The normal probability model is shown with the dashed line.   199

9.9   Q-Q plot of the bootstrapped scores drawn in the first replicate. The line references where the points would lie if the sample of scores was consistent with a normal distribution.   201

9.10   Q-Q plot of the bootstrapped standardized scores for $R = 25$ replicates of $n = 116$ resampled observations from the standard normal distribution. The bold line references where the points would lie if the sample of scores was consistent with a normal distribution.   202

9.11   Q-Q plot of the studentized Mexican achievement scores along with a 95% confidence envelope. The confidence envelope was computed using $R = 4999$ replicates of $n = 116$ observations resampled using a parametric bootstrap from the $N(0, 1)$ distribution.   203

9.12   Density plot of the empirical standardized Mexican achievement scores. The confidence envelope based on the normal probability model is also added to the plot.   203

10.1   Distribution of PSAT scores conditioned on treatment.   209

10.2   Density plots for the distribution of mathematics achievement scores conditioned on treatment.   211

10.3   Side-by-side box-and-whiskers plots for the distribution of mathematics achievement scores conditioned on treatment.   211

10.4   Randomization distribution of the mean difference under the null hypothesis of no difference. A vertical line (solid) is drawn at 0, the hypothesized mean difference.   216

10.5   Randomization distribution of the mean difference for both the assumptions of independence and dependence. A vertical line (dotted) is drawn at the hypothesized mean difference of 0.   219

10.6   The steps used to block bootstrap the distribution for a particular test statistic, $\hat{V}$, under the alternative model.   220

10.7   Bootstrap distribution of the standardized mean difference as measured by Glass's delta in mathematics achievement scores between treatment and control participants for 4999 replicates. The resampling was carried out using block bootstrapping under the alternative model. The vertical line (dotted) is drawn at the observed standardized mean difference of 0.51.   223

11.1   Conditional density plots showing the distribution of weight loss for the Atkins and the Ornish diets.   231

11.2   Conditional density plots showing the distribution of weight loss for the Atkins and the other three diets.   233

11.3   Conditional density plots showing the distribution of weight loss for the Atkins/Zone and the LEARN/Ornish diets.   235

11.4   Steps used to carry out a randomization test to obtain the distribution for a particular contrast under the null model of no difference.   239

11.5   Side-by-side box-and-whiskers plots showing the distribution of word recall conditioned on time spent studying. The conditional means are each marked with a triangle. The dotted line, which connects the conditional means, shows a potential polynomial trend. 249

12.1   Side-by-side box-and-whiskers plots of 12-month weight change conditioned on diet.   256

12.2   The deviation between the observed score and the marginal mean (horizontal line) for four observations.   263

12.3   The deviation between the observed score and the marginal mean (horizontal line) for four observations. Each has now been decomposed into the part of the deviation that can be explained by diet (solid) and that which cannot (dashed).   264

12.4   Graphical representation of the squared deviations from the marginal mean weight change for four observations.   265

12.5   Graphical representation of the squared decomposed model deviations for four observations—the deviation between the conditional mean (dotted horizontal line) and the marginal mean (solid horizontal line). The squares represent the squared model deviations for the four observations.   266

12.6   Graphical representation of the squared decomposed residual deviations for four observations—the deviation between the conditional mean (dotted horizontal line) and the observed value. The squares represent the squared residual deviations for the four observations.   267

12.7   Steps to carry out a randomization test to obtain the distribution for the omnibus effect under the null hypothesis.   268

12.8   Dynamite plot showing the conditional mean weight change and standard error for each diet.   278

12.9   Plot showing the point estimate and bootstrap interval estimate of the mean difference for each contrast based on the bias-corrected-and-accelerated adjustment using 4999 bootstrap replicates.   280

12.10   Plot showing the point estimate and bootstrap interval estimate of the mean difference for each contrast based on the Benjamani–Hochberg adjusted significance level. Each interval uses the bias-corrected-and-accelerated adjustment using 4999 bootstrap replicates. A vertical line is plotted at zero, the parameter value associated with no difference.   283

LIST OF TABLES

2.1	Comparison Operators Used in R	38
2.2	Logical Operators Used in R	39
4.1	Common Colors and Their Associated RGB Values Using a Maximum Color Value of 255	83
4.2	Small Data Set (original data) Showing Results of Trimming (trimmed data) and Winsorizing (winsorized data) Distribution by 20%	88
4.3	Comparison of Conventional and Robust Estimates of Location and Variation for Conditional Distributions of Household per Capita Expenditures	91
5.1	Mean, Standard Deviation and Sample Sizes for Household per Capita Expenditures (in dollars) Conditioned on Region[a]	103
5.2	Mean, Standard Deviation and Sample Sizes for Household per Capita Expenditures (in dollars) Conditioned on Urbanicity and Region	105
6.1	Four Potential Scenarios Researcher Could Face When Making Inferences	117

6.2   Ten Possible Permutations of the Perceived Social Competence Scores and Difference in Mean Perceived Social Competence Scores   121

7.1   Probability Distribution of Random Variable X for Number of Heads When Tossing 10 Coins   142

7.2   Results of Carrying Out Nonparametric Bootstrap Test for Three Different Test Statistics[a]   164

8.1   Four Potential Scenarios Researcher Could Face When Making Inferences[a]   173

8.2   Fisher's Descriptions for Scale of Evidence Against the Null Hypothesis[a]   177

9.1   Parameter Value, p-Value, and Qualitative Description of Strength of Evidence Against the Model for 11 Potential Models   185

10.1   Data for Participants in First Two Blocks Presented in Long Format   212

10.2   Data for Participants in First Two Blocks Presented in Wide Format   213

12.1   Mean and Standard Deviation for 12-month Weight Change (in kg) Conditioned on Diet.[a]   256

12.2   Contrast Coefficients, Observed Contrast Value, Randomization Test Results, Eta-Squared Estimate, and Bootstrap Results for Interval Estimate of Eta-Squared for Each of Six Pairwise Contrasts[a]   271

12.3   Randomization Test Results for Both Unadjusted Method and Benjamani–Hochberg Ensemble-Adjusted Method[a]   276

12.4   Computation of Adjusted Significance Level for Benjamani–Hochberg Simultaneous Confidence Intervals Based on Sequential Unadjusted p-Values   281

12.5   Unadjusted and Simultaneous Bootstrap Intervals for Each of Six Contrasts[a]   282

FOREWORD

This book is truly different.

It is rare that a textbook with so few technical prerequisites offers so much of value, not just to its intended student audience, but also to all practicing statisticians. The book in your hands is one of those rare ones. It is the product—over years—of careful study and deep thought by its authors, who are practicing statisticians, education researchers, and root-deep innovators.

Statistics, its teaching and its learning, have seen a number of exciting developments over the last quarter century. Among these:

1. Near-universal acceptance of the assertion that applied statistics textbooks should emphasize the priority of real data over mathematically driven abstract exposition.

2. A coming-of-age for computer-intensive methods of data analysis.

3. A parallel recognition by teachers of statistics that, along with an emphasis on real data and reliance on applied context, computer simulation offers an alternative to abstract derivations as an approach to exposition and understanding.

4. A cooperative development and broad embrace of R, the public domain, and open-source software, whose growing community of contributors has made cutting edge, computationally intensive methods freely available around the world.

5. A coming-of-age of statistics education as a research area in its own right, and, as a result, a growing clarity about what is hardest for students to learn, and a sharpened focus on what is most important for students to learn.

6. A change at the foundations of our subject: What we can compute shapes what we can do; what we can do shapes what we actually do in practice; what we do in practice shapes how we think about what we do. (As just one example, when I first was learning statistics in the 1960s, Bayesian analyses were typically dismissed as mere idealism; today, the combination of hierarchical models and Markov chain Monte Carlo has turned Bayes resistors into old fogeys.)

By integrating these developments in a novel way, *Comparing Groups* deserves to be regarded as one of this quarter-century's pioneering textbooks aimed at introducing nonstatisticians to contemporary thought and practice.

Many features are distinctive:

1. The exposition is visual/intuitive/computational rather than haunted by the old headless horseman of abstract derivations.

2. The writing is strikingly good—the exposition reflects the authors' careful attention to word choice. (In my experience, it is rare in statistical exposition to read with a sense that the writers made thoughtful, deliberate choices about wording after considering a variety of options.)

3. The references are a gold mine.

4. The use of modern graphics is unusual for a book at this level. I refer, in particular, to panel plots and kernel density estimates.

5. Prerequisites are minimal. No calculus is needed. Although it helps to know about vectors and matrices, one does not need a linear algebra course.

6. The emphasis is practical throughout. For example, the exposition refers systematically to APA guidelines and offers sample write-ups.

7. Content reflects the current research literature. For example, the exposition recognizes the importance of effect sizes, and the treatment of multiple testing addresses the recent research on false discovery rates.

8. Overall, the emphasis is on statistics with a purpose, statistics for deep interpretive understanding.

In short, this is a book that reduces prerequisites, avoids technical baggage, focuses on essentials, and teaches via authentic applications. It offers readers our profession's current best sense of how to understand data for comparing groups, taking full advantage of computationally intensive methods such as randomization and the bootstrap.

George W. Cobb
Robert L. Rooke Professor of Statistics, emeritus, Mount Holyoke College
Vernon Wilson Endowed Visiting Professor, Eastern Kentucky University

PREFACE

*Computational advances have changed the face of statistical practice by transforming what
we do and by challenging how we think about scientific problems.*
—R. A. Thisted & P. F. Velleman (1992)

Drawing conclusions and inferences about the differences among groups is an almost
daily occurrence in the scholarly life of an educational or behavioral researcher.
Group comparisons are at the heart of many research questions addressed by these
researchers. These are questions about efficacy, such as "Is a particular curriculum
effective in improving students' achievement?" and questions about magnitude such
as "How much lower are attendance rates for a particular population of students?"

The content in this book provides the statistical foundation for researchers interested in answering these types of questions through the introduction and application
of current statistical methods made possible through computation—including the use
of Monte Carlo simulation, bootstrapping, and randomization tests. Rather than focus
on mathematical calculations like so many other introductory texts in the behavioral
sciences, the approach taken here is to focus on conceptual explanations and the use
of statistical computing. We agree with the sentiments of Moore (1990, p. 100),
who stated, "calculating sums of squares by hand does not increase understanding; it
merely numbs the mind."

At the heart of every chapter there is an emphasis on the direct link between research questions and data analysis. Purposeful attention is paid to the integration of design, statistical methodology, and computation to propose answers to research questions based on appropriate analysis and interpretation of quantitative data. Practical suggestions for analysis and the presentation of results based on suggestions from the *APA Publication Manual* are also included. These suggestions are intended to help researchers clearly communicate the results of a data analysis to their audience.

Computation as a Tool for Research

Computation is ubiquitous in everyday life. This is in large part due to the progress made in technology in the last 20 years. The price of computational power continually becomes less expensive and more powerful. In a recent *New York Times* article Caleb Chung—the inventor of the electronic Furby toy—was quoted as saying, "the price of processing power has dropped to the floor. I can buy the equivalent of an Apple II processor for a dime" (Marriot, 2007, p. 9).

Computing has become an essential part of the day-to-day practice of statistical work. It has not only greatly changed the practice of statistics itself, but has influenced the development of new state-of-the-art statistical methodologies and broadened the types of questions that can now be addressed by research scientists applying these newly derived data analytic techniques. Previous to having the availability of such computational power, the outcome of a research project may have been crucially dependent on the efficiency or numerical accuracy of algorithms employed by a methodologist. Computational advantages have allowed educational and behavioral researchers to capitalize on sheer computing power to solve problems that previously were impractible or intractable.

Although computer-based data analysis probably covers most of the activity that educational and behavioral researchers use in their work, statistical computing is more than just using a software package for the analysis of data. It also encompasses the programming and development of new functionality or software, the analysis of statistical and numerical algorithms, and data management. Furthermore, computing is an important avenue for allowing statisticians and scientists to pursue lines of inquiry that in the past would not have been possible. It is also an essential tool in any modern collaborative effort.

To support and help facilitate the use of scientific computing, examples using the R computer language will be used throughout the book. Rather than relegate examples to the end of chapters, our approach is to interweave the computer examples with the narrative of the monograph. R is free, which means it is available to individuals with various resources. It is also a professional-level software environment that implements many of the modern statistical methods emphasized in the monograph. Furthermore, the architecture of the object orientation in R easily allows data analysis to be performed based on structured inquiry, in which a series of interrelated and increasingly complex questions are posed to guide and inform the process of data analysis. In the classroom, structured inquiry can increase the opportunities for intellectual engagement (e.g., Badley, 2002; Cooper, 2005; Lee, 2005; Prince & Felder, 2006). Lastly, since R is a functional programming language, it can be easily extended for use in a great many analytic scenarios in the behavioral and educational sciences.

Organization of the Book

The organization of this book has been shaped by the authors' experiences in teaching the statistical and computing content presented in many graduate courses for social science students. The topics introduced represent the authors' beliefs about relevant content for inclusion in an introductory graduate-level statistics course for educational and behavioral science students.

This content covers statistical computing, exploratory data analysis, and statistical inference. The individual chapters in the book endeavor to integrate these ideas to help guide educational and behavioral researchers through the process of data analysis when making group comparisons. While most chapters address multiple strands of the data analytic process, in surveying the computing and statistical content of each chapter, it is clear that each chapter is primarily focused to address a particular strand.

Statistical Computing

The first two chapters are written primarily for educational and behavioral researchers who may not have prior experience with computing environments and, in particular, with the R statistical computing environment. Chapter 1 introduces the fundamentals of R. Basic ideas of computation are presented as the building blocks for the remainder of the monograph. In Chapter 2, these ideas are further developed through the reading in and manipulation of external data (data saved outside of R). This chapter also introduces script files as a way of recording useful syntax.

Exploratory Data Analysis

Chapters 3, 4, and 5 introduce graphical and numerical methods of exploratory analysis for group comparisons. These chapters begin the integration of the statistical and computing content. Chapter 3 focuses on graphical exploration of a marginal distribution to introduce the statistical and computing content without the additional cognitive load of making comparisons. Ideas of kernel density estimation are at the forefront of this chapter. Additionally, the use of R to create publication quality plots is highlighted.

Chapters 4 and 5 focus on the exploration of conditional distributions for two and more than two groups, respectively. Graphical exploration is expanded to facilitate comparisons (e.g., panel plots) and numerical summary measures are introduced to quantify characteristics of the distributions.

Statistical Inference

Chapter 6 is the first of two chapters on the use of Monte Carlo methods for statistical inference. This chapter presents randomization and permutation tests to examine group differences. Chapter 7, the second such chapter, presents parametric and nonparametric bootstrap tests. Both of these chapters highlight the quantification of how likely an observed statistic is given the expected variation in that statistic because of chance. Chapter 6 highlights the expected random variation due to random assignment, whereas Chapter 7 highlights the random variation due to random sampling.

Chapter 8 expands on the differences between the randomization and bootstrap tests by offering philosophical reasons for using each method apart from the design employed by the researcher. This is especially highlighted for observational studies. This chapter also offers an overview of differences between the Fisher and Neyman–Pearson philosophies of statistical testing.

Chapters 9 and 10 expand the realm of statistical inference. Chapter 9 introduces ideas of interval estimation. The bootstrap methodology is developed further in this chapter, but under stratification of groups. Standardized and unstandardized measures of effect size are discussed. In Chapter 10, testing and estimation for dependent samples is introduced. Rather than do this through the use of repeated measures, which is often the case in statistics books in the educational and behavioral sciences, dependence is introduced via the use of blocking in the research design.

Chapters 11 and 12 introduce common methods in the educational and behavioral sciences for comparing more than two groups. Chapter 11 is focused on planned comparisons. Exploration, testing and estimation are revisited in the context of multiple groups. The use of linear contrasts is introduced to facilitate these ideas.

Chapter 12 is focused on unplanned comparisons. Three common methods are presented: (1) adjusted group comparisons without the omnibus test; (2) unadjusted group comparisons following an omnibus test; and (3) adjusted group comparisons following an omnibus test. Strengths, weaknesses, and criticisms of each method are discussed. The use of ensemble-adjusted p-values and adjusted bootstrap intervals are discussed in this chapter.

Extras

This book refers to and uses several data sets throughout the text. Each of these data sets is available online at http://www.tc.umn.edu/~zief0002/ComparingGroups html. The codebook for each data set is also available at the Web site. Many of these data sets were used in actual research studies that have been published and were graciously provided by the authors and researchers involved in those studies.

The R script files for each chapter are also available at the URL mentioned above. The commands are provided so that the reader can reproduce all the output and plots discussed in the monograph.

Problems can be found at the end of each chapter, except Chapter 8. These problems are meant to provide a platform to perform data analyses using real data which integrates and extends the main ideas from each chapter. We provide exemplar solutions for each question as well as the R script necessary to carry out all necessary computations.

In the book, there are several typographic conventions that are used. Text in the book that appears in `typewriter text` indicates R commands or syntax. **Bold-faced text** indicates a particular R package (see Chapter 1).

ACKNOWLEDGMENTS

The authors wish to express their thanks to Christopher Gardner, Denise Gottfredson, Nguyen Phong, Katherine Stamps Mitchell, and all of their co-authors for generously sharing data from their research studies. A special thank you also is given to George Cobb who agreed to write the foreword. The authors are also indebted to the reviewers who provided invaluable feedback that helped improve this manuscript.

Thanks to my wife for endless support and to my dogs for forcing me to take the time to stop and scratch their bellies. A very special thank you goes to Joan Garfield, whose continued support and encouragement not only got me started on this journey but saw it through to the end. Thanks also to my family (Mom, Dad, and Bhone), friends (Tim, Heather, Charles, and many other), and students (especially Julio, Laura, Rebekah, and Stacy) who put up with me during the last year. And a very special thank you to Grace, Rex, Teresa, and Tiera.

A. S. Z.

Thanks to Joan, Adam, and Emma for continuing to be my inspiration.

J. R. H.

Thanks to Leslie for her patience and encouragement.

J. D.L.

CHAPTER 1

AN INTRODUCTION TO R

Computing is an essential component of statistical practice and research. Computational tools form the basis for virtually all applied statistics. Investigations in visualization, model assessment and model fitting all rely on computation.

—R. Gentleman (2004)

R is a rich software environment for statistical analysis. Hardly anyone can master the whole thing. This chapter introduces the most basic aspects of R, such as installing the system on a computer, command and syntax structure, and some of the more common data structures. These ideas and capabilities will be further developed and expanded upon in subsequent chapters as facility is built with using R for data analysis. Lastly, this chapter introduces some of the practices that many R users feel are instrumental, such as documenting the data analysis process through the use of script files and syntactic comments.

Comparing Groups: Randomization and Bootstrap Methods Using R
First Edition. By Andrew S. Zieffler, Jeffrey R. Harring, & Jeffrey D. Long
Copyright © 2011 John Wiley & Sons, Inc.

1.1 GETTING STARTED

This section introduces some of the initial steps needed to get started using R. It includes steps for downloading and installing the R system on a computer and instructions on how to download and install add-on packages for R.

R can be downloaded and installed from the CRAN (Comprehensive R Archive Network) website at `http://cran.r-project.org/`. Click on the appropriate link for the operating system—Windows, MacOS X, or Linux—and follow the directions. At the website, the precompiled binary rather than the source code should be selected. Further instructions are provided below for the two most common operating systems, Windows and Mac.

1.1.1 Windows OS

Click on the base link and then click `Download R 2.12.0 for Windows` (or whatever the latest version happens to be). Accept all the default options for installation. After installation, an R icon will appear on the desktop. Double-click the icon to start the program. If the software is successfully downloaded and installed, the opening screen should look something like Figure 1.1.

1.1.2 Mac OS

Click the link for `R-2.10.1.dmg` to download (or whatever the latest version happens to be). To install R, double-click on the icon of the multi-package `R.mpkg` contained in the `R-2.10.1.dmg` disk image. Accept all the defaults in the installation process. If installed successfully, an R icon will be created, typically in the *Applications* folder. Double-click the icon to start the program. If the software is successfully downloaded and installed, the opening screen should look like Figure 1.1.

1.1.3 Add-On Packages

R functions and data sets are stored in *packages*.[1] The basic statistical functions that researchers in the educational and behavioral sciences use are part of packages that are included in the base R system. The base system is part of the default R installation. Other useful functions are included in packages that are not a part of the basic installation. These packages, which often include specialized statistical functionality, are contributed to the R community and can often be downloaded from CRAN directly or installed from within R.

To install a package, the `install.packages()` function is used. If this is the first time a package is installed, after executing `install.packages()`, a list of all of the CRAN mirror sites will be presented. After selecting a mirror site, a list of available packages will appear. (This is the starting point after the first package installation.) Select the appropriate package desired and R will download and install it.

---

[1]Some authors and instructors will use the term *library* instead of *package*.

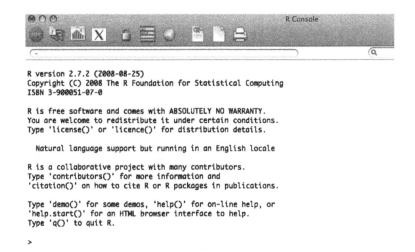

```
R version 2.7.2 (2008-08-25)
Copyright (C) 2008 The R Foundation for Statistical Computing
ISBN 3-900051-07-0

R is free software and comes with ABSOLUTELY NO WARRANTY.
You are welcome to redistribute it under certain conditions.
Type 'license()' or 'licence()' for distribution details.

  Natural language support but running in an English locale

R is a collaborative project with many contributors.
Type 'contributors()' for more information and
'citation()' on how to cite R or R packages in publications.

Type 'demo()' for some demos, 'help()' for on-line help, or
'help.start()' for an HTML browser interface to help.
Type 'q()' to quit R.

>
```

Figure 1.1: Example of the R console window on a Mac.

The name of the package can also be typed—inside of quotation marks—directly into the `install.packages()` function. Command Snippet 1.1 shows the syntax for downloading and installing the **sm** package which contains functions for smoothing, which is used in Chapter 3. The command snippet shows the optional argument `dependencies=TRUE`. This argument will cause R to download and install any other packages that the **sm** package might require.

Command Snippet 1.1: Syntax to download and install the **sm** package.

```
> install.packages(pkgs = "sm", dependencies = TRUE)
```

Many package authors periodically update the functionality in their packages. They may fix bugs or add other functions or options. The `update.packages()` function is used to update any packages that have been installed. Such updating should be done periodically, say, every few months.

Installing a package downloads that package and installs it to the R library. To use the functions that exist in these installed packages, the package needs to be loaded into the R session using the `library()` or `required()` function. Command Snippet 1.2 shows the syntax to load the **sm** package into an R session. After successfully loading the package, all of the functions and data sets in that package are available to use. The package will not need to be loaded again during the R session. However, if the R session is terminated, the package must be loaded in the new session.

A multitude of add-on packages, called *contributed packages*, are available from CRAN (see http://cran.r-project.org/web/packages/). Additional add-on packages are available through other package repositories. For example, *The Omega*

Command Snippet 1.2: Loading the **sm** package.

```
> library(sm)
```

Project for Statistical Computing (see http://www.omegahat.org/) includes a variety of open-source packages for statistical applications, particularly for web-based development. *Bioconductor* (see http://www.bioconductor.org/) is a repository of packages for the analysis and comprehension of genomic data. A repository of packages that are available, but still under development, is located at *R-Forge* (see http://r-forge.r-project.org/).

In general, the argument repos= is added to the install.packages() function to specify the URL associated with the package repository. For example, Command Snippet 1.3 shows the syntax to download and install the **WRS** package from the *R-Forge* repository. The websites for each of the repositories has more specific directions for downloading and installing available add-on packages.

Command Snippet 1.3: Installing the **WRS** package from the R-Forge repository.

```
> install.packages(pkgs = "WRS", dependencies = TRUE, repos =
    "http://R-Forge.R-project.org/")
```

In addition to the **sm** and **WRS** packages, the add-on packages **colorspace, dichromat, e1071, MBESS, quantreg**, and **RColorBrewer** are used in this monograph.

1.2 ARITHMETIC: R AS A CALCULATOR

R can be used as a standard calulator. The notation for arithmetic is straightforward and usually mimics standard algebraic notation. Examples are provided in Command Snippet 1.4.

The character > in the R terminal window is called a *prompt*. It appears automatically and does not need to be typed. The [1] indicates the position of the first response on that line. In these simple cases, when there is only one response, the [1] seems superfluous, but when output spans across several lines, this provides a very useful orientation. Note the special value Inf returned for the computation 1/0. There are three such special values: Inf, -Inf, and NaN. The first indicates positive infinity, the second indicates negative infinity, and the third means the result is not a number.

1.3 COMPUTATIONS IN R: FUNCTIONS

There are many buit-in functions for performing mathematical and statistical computations in R. There are three basic components for any computation.

Command Snippet 1.4: Examples of arithmetic computations.

```
## Addition
> 3 + 2
[1] 5

## Multiplication
> 4 * 5
[1] 20

## Exponentiation
> 10 ^ 3
[1] 1000

## Division
> 1 / 0
[1] Inf
```

- Function: e.g., sqrt(), log(), cos(), ...

- Arguments: The inputs to the function

- Returned Value: The output from the function applied to the arguments

Several examples of computations are given in Command Snippet 1.5. Examine the first line of Command Snippet 1.5. The components of this computation are

$$> \quad \overbrace{\text{sqrt}(\ \underbrace{x = 100}_{Argument}\ )}^{Function}$$

$$[1] \quad 10 \qquad \leftarrow Returned\,Value$$

Notice that the returned value of a computation is indexed in the same way as the returned value of arithmetic computations. In these initial examples, there is one argument that is unnamed. Often in data analysis, more complex computations are warranted. As more complex computations are used, there are a few simple rules to observe. These are listed below and illustrated in Command Snippet 1.6.

- Argument(s) are always enclosed in parentheses.

- When there are multiple arguments, they are separated by commas.

- Many functions accept optional "named arguments" that specify some aspect of the computation.

- The order of named arguments doesn't matter, but the order of the other arguments does.

Command Snippet 1.5: Examples of computations using functions.

```
## Square root
> sqrt(100)
[1] 10

## Natural logarithm
> log(7)
[1] 1.94591

## Sine of angle given in radians
> sin(50)
[1] -0.2623749

Exponentiation base e
> exp(3)
[1] 20.08554
```

Command Snippet 1.6: Examples illustrating how arguments are provided in functions.

```
## Two unnamed arguments
> log(100, 10)
[1] 2

## Two unnamed arguments in reverse order
> log(10, 100)
[1] 0.5

## Two named arguments
> log(x = 100, base = 10)
[1] 2

## Two named arguments in reverse order
> log(base = 10, x = 100)
[1] 0.5

## First argument is unnamed and subsequent arguments are named
> log(100, base = 10)
[1] 2
```

Many R users leave the first argument unnamed and name all subsequent arguments. The precedent for most examples in the remainder of this monograph will be to leave the first argument unnamed, and name all subsequent arguments in a function.

Sometimes a computation does not make sense and R will produce an error statement. Other times a computation is odd in some way—as judged by the people who wrote the software—and a value is returned, but there is also a warning message. Recall that NaN is a special numerical value. It means "Not a number."[2]

For novice users, the error statements and warnings may seem cryptic. However, as familiarity with the language is developed, reading the error statement can often help a data analyst figure out what is going wrong. It is often helpful to include the error statement or warning message along with the code if one is seeking help. Command Snippet 1.7 shows an example of both an error and warning message.

Command Snippet 1.7: Errors and warnings in computations.

```
## Error in computation
> log(X)
Error: object 'X' not found

## Warning message
> sqrt(-3)
[1] NaN
```

1.4 CONNECTING COMPUTATIONS

One of the advantages of R is that the return value from one computation can be taken as the input to another computation. This is very helpful for performing successive operations and for accessing important aspects of statistical output. For example, suppose the goal is to find the natural logarithm of a number and then take the square root of the result. There are two basic styles for doing this, *chaining* and *assignment*.

Chaining computations together uses one, or more, computations directly as the argument(s) in another computation. Command Snippet 1.8 shows the use of chaining to connect computations.

Command Snippet 1.8: Connecting computations through chaining.

```
## Find the sine of pi/2, and then take the square root of the
     result
> sqrt(sin(pi / 2))
[1] 1

## Find the base 10 logarithm of 100, and then take the square
     root of the result
> sqrt(log(100, base = 10))
[1] 1.414214
```

The second way of connecting computations uses assignment. The returned value of a computation can be stored by assigning it to a named object. (Think of it as

---

[2]For the mathematically inclined, the computation of $\sqrt{-3}$, for example, involves complex numbers. R will perform complex arithmetic. For example, to compute $\sqrt{-3}$ we use sqrt(-3+0i).

storing the result of a computation into a variable.) Then, the named object can be passed to the subsequent computation. The assignment operator is "<-" constructed by using the "<" key followed by the "-" key. Command Snippet 1.9 shows the use of assignment to connect computations.

Command Snippet 1.9: Connecting computations through assignment.

```
## Assign the base 10 logarithm of 100
> chili <- log(100, base = 10)

## Find the square root
> sqrt(chili)
[1] 1.414214

## Reassign the object name
> chili <- 3

## Find the square root
> sqrt(chili)
[1] 1.732051

## Reassign the object name
> chili <- 25

## Find the square root
> sqrt(chili)
[1] 5

## Print the value of the object
> chili
[1] 25

## List all objects assigned in the session
> ls()
[1] "chili"
```

When a name is reused to assign a different object, the previous value of that object is lost irretrievably. To see the value associated with an object, use the name as if it were a command in R. All currently assigned objects can be viewed by issuing the list function, ls(), with no arguments.

1.4.1 Naming Conventions

When naming objects in R, there are a few rules to abide by.

- Names can only include letters, digits, and periods.

- Names cannot begin with a digit, a period followed by a digit, or a special character (e.g., #).

- Some names should be avoided since they already have special meaning given to them by R. For example, TRUE and FALSE, or their shortened versions T and F.

Aside from the above rules, object names are fairly open. It is good practice to name objects so that they are descriptive of what they contain. For example, storing the number 25 in chili is not descriptive of what the object contains. A better name, describing the current contents might be n25. An even better alternative might be number25, or even thenumbertwentyfive, though the longer the name the more typing involved.

There are a number of conventional ways to create object names without spaces that are easier for humans to read. One of those conventions is to use *bumpy case*. Bumpy case combines upper- and lowercase letters to break up the different words like TheNumberTwentyFive. As you look at that name, you will probably be able to tell why it is referred to as bumpy case. Another naming convention—and the one that will be used throughout the remainder of this monograph—is to replace spaces with periods, as in the.number.twenty.five.

1.5 DATA STRUCTURES: VECTORS

One of the most fundamental data structures used in R is the *vector*. A vector is a unidimensional array (arrangement) of values, either a row or a column. Vectors are a nice way to display data from a single variable, known as *univariate* data. For example, consider a vector of ages for five children arrayed as a column vector:

$$
\text{Ages} = \begin{pmatrix} 1 \\ 6 \\ 7 \\ 7 \\ 10 \end{pmatrix}.
$$

1.5.1 Creating Vectors in R

Short vectors, like the five children's ages above, can be entered directly into R by "collecting" the data into a vector using the c() function. The c is short for *concatenate* which simply appends each subsequent argument provided in the c() function into a single vector. The vector is typically assigned to an object so that computations can be performed on the data in the collection. Command Snippet 1.10 shows the syntax to create the vector of ages in the example above and assign it to an object called ages.

While very useful, using the c() function is not the only manner in which a vector can be constructed. Two other functions commonly used are seq() and rep(). The seq() function produces a sequence of values using the arguments from=, to=, and by=. For example, to create a vector containing elements that consist of the even numbers from 2 to 24, any of the computations in Command Snippet 1.11 could be employed.

Command Snippet 1.10: The c() function is used to make a collection.

```
## Assign the vector of ages to an object
> ages <- c(1, 6, 7, 7, 10)

## Print the object
> ages
[1]  1  6  7  7 10
```

Command Snippet 1.11: Examples of creating a vector of the even numbers from 2 to 24 using the c() function and the seq() function.

```
## Using the c() function to create a vector
> X <- c(2, 4, 6, 8, 10, 12, 14, 16, 18, 20, 22, 24)
> X
 [1]  2  4  6  8 10 12 14 16 18 20 22 24

## Using the seq() function to create the same vector
> X <- seq(from = 2, to = 24, by = 2)
> X
 [1]  2  4  6  8 10 12 14 16 18 20 22 24
```

When by=1, a shortcut is to use the colon operator (:). The colon operator can be inserted between two values to produce the sequence having steps of 1 (e.g., $= 12, 13, 14, \ldots, 23, 24$). When using the colon operator to create a sequence, neither the c() nor seq() functions need be used. For example, to create a vector of the sequential values from 1 to 10, the syntax in Command Snippet 1.12 is used.

Command Snippet 1.12: Examples of creating a vector of the sequential values from 1 to 10 using the colon operator and the seq() function.

```
## Sequence using the colon operator
> X <- 1:10
> X
 [1]  1  2  3  4  5  6  7  8  9 10

## Same sequence using seq()
> X <- seq(from = 1, to = 10, by = 1)
> X
 [1]  1  2  3  4  5  6  7  8  9 10
```

The rep() function is used to create vectors of repeated values in R. The first argument to this function, x=, is the value to be repeated. The argument times= takes a value indicating the number of times the first argument is repeated. For example, say the goal is to create a vector composed of 10 elements where each element is the value 1. Command Snippet 1.13 shows the syntax for creating this vector using the rep() function.

Command Snippet 1.13: Example of creating a vector composed of 10 elements where each element is the value 1 using the rep() function.

```
> X <- rep(1, times = 10)
> X
 [1] 1 1 1 1 1 1 1 1 1 1
```

The arguments x= and times= can also be collections, or vectors, of elements. For example, if the object is to create a vector where the first 10 elements are the value 1 and the next 15 elements are the value 0, the syntax in Command Snippet 1.14 is used.

Command Snippet 1.14: Example of creating a vector composed of 25 elements where the first 10 elements are the value 1 and the next 15 elements are the value 0.

```
> X <- rep(c(1, 0), times = c(10, 15))
> X
 [1] 1 1 1 1 1 1 1 1 1 1 0 0 0 0 0 0 0 0 0 0 0 0 0 0 0
```

1.5.2 Computation with Vectors

Vectors can be used as an argument to many R functions or in arithmetic computations. Some functions deal with vectors by applying the particular computation to each element of the collection. Other functions combine the elements of the collection in some way before applying the computation. Functions of both types are shown in Command Snippet 1.15.

Command Snippet 1.15: Different functions applied to a collection of data.

```
## Add five to each element
> ages + 5
[1]  6 11 12 12 15

## Averages all of the elements
> mean(ages)
[1]  6.2
```

The functions available in R for performing computations on vectors are too numerous to be listed here. R includes functions for several basic mathematical

functions (log(), exp(), log10(), cos(), sin(), sqrt(),...) as well as many
functions that are especially useful in statistics. In addition, there are other functions
that perform computations on vectors such as sort(), rev(), order(), rank(),
scale(), etc. that return more complex results. Each of these functions will be
discussed in detail when the need arises.

1.5.3 Character and Logical Vectors

The vectors created thus far are all numerical vectors—all of the elements are numbers. There are two other common types of vectors used in R. One of these is a
character vector. The elements of character vectors are *character strings* or *literals*.
Character strings are just a sequence of characters demarcated by quotation marks.
When elements are inside of quotation marks, this tells R not to look for a value.
Categorical data are often stored as character strings or as a closely related form
known as a *factor*. Command Snippet 1.16 shows an example creating a character
vector and assigning that vector to an object called educational.level.

Command Snippet 1.16: Examples of character and logical vectors.

```
## Character vector
> educational.level <- c("High School", "College", "Some
    College", "College", "College" )

## Logical vector
> logical.vector <- c(FALSE, TRUE, FALSE, TRUE, TRUE)

## Conditional statement generating a logical vector
> ages > 5
[1] FALSE  TRUE   TRUE   TRUE   TRUE

## Assigning the logical vector to an object
> older.than.five <- ages > 5
> older.than.five
[1] FALSE  TRUE   TRUE   TRUE   TRUE
```

Another type of vector that R supports is a *logical vector*. The elements in a logical
vector are either TRUE or FALSE (R requires these be in all uppercase). To differentiate
logical vectors from character vectors, there are no quotes around TRUE and FALSE
when they are used as values. Command Snippet 1.16 also shows an example creating
a logical vector and assigning that vector to an object called logical.vector.

While the c() function can be used to create a logical vector, they are more
often generated through the use of conditional statements. For example, a researcher
might want to keep track of which values of ages are greater than 5. She could
type the logical values for each element into a new collection. This method is
shown in Command Snippet 1.16 when assigning the logical values in the object
logical.vector. However, it is quicker to write a *conditional statement* and then
use the assignment operator to store the logical values produced into a new vector.

This method is also shown in Command Snippet 1.16 in the assignment of the object `older.than.five`.

1.6 GETTING HELP

The R Development Core Team has written several helpful documents for the novice user, such as *The R FAQ* (Hornik, 2010), *An Introduction to R* (Venables, Smith, & the R Development Core Team, 2009), and *R Data Import/Export* (R Development Core Team, 2009). These documents, along with many others, are available as PDF files on the R website (`http://cran.r-project.org/Documentation/Manuals`) or directly via the Help menu within R.

The R program installation includes extensive documentation for all the functions. These can be accessed at any time by typing `help(`*function*`)` or `?`*function*, where *function* is the particular function for which help is sought (e.g., `help(cor)`).

There are two things to note about this documentation. First, *function* needs to be an actual function that R recognizes. Secondly, the top portion of the help page usually consists of options and usage operators of the function. The bottom portion of the help page usually provides examples on the use of the function with particular options and operators. As an example of the first point, suppose `help(correlation)` is executed rather than `help(cor)`. This will produce a pop-up window suggesting that R cannot find the help documentation for that function. That is because there is no function called `correlation`.

The `help.search()` function or `??` can be useful when the exact name of a function is unknown. Each will search for strings of characters in all of the help documentation that is available on the R system. For example, `??correlation`, will look for the word "correlation" in the help documents. Depending on which packages have been installed (see Section 1.1.3), there may be several potential functions which could be used to compute the correlation. One of these is `cor()`. Thus, the `help()` function needs to be given `cor` as its argument rather than `correlation`. Command Snippet 1.17 shows the syntax for using both `?` and `??`.

Command Snippet 1.17: Accessing the help documentation in R.

```
## Documentation for the cor() function
> ?cor

## Produces an error
> help(correlation)
No documentation for 'correlation' in specified packages and
    libraries:
you could try '??correlation'

## String search for "correlation"
> ??correlation
```

This documentation is quite useful for experienced users of R, but can be frustrating for those who are new to R. Reading the documentation when first learning R is a little like trying to learn English by studying the *Oxford English Dictionary*. But, once a little of the language is learned, it becomes easier to understand the object of the dictionary. For novice users, one of the best ways to find help or more readable documentation is to do a Google search. As R is becoming more popular, many documents, course notes, and books are being made readily available online.

1.7   ALTERNATIVE WAYS TO RUN R

The examples above have all involved a console-based graphical user interface(GUI) to run R interactively, in which a command is typed and then executed by R when the Enter key is pressed. There are additional GUI-based options available that provide a menu system with point-and-click options for carrying out operations. The most mature of these options to date is the **Rcmdr** package (R Commander; http://cran. r-project.org/doc/packages/Rcmdr.pdf.). **Rcmdr** must be downloaded and installed like any other package. However, when the syntax library(Rcmdr) is issued, the standard R console is replaced by the R Commander console. While a point-and-click system might sound better than typing at the command line, as John Fox, the author and programmer of *R Commander*, points out:

> In my opinion, a GUI for statistical software is a mixed blessing: On the one hand, a GUI does not require that the user remember the names and arguments of commands, and decreases the chances of syntax and typing errors. These characteristics make GUIs particularly attractive for introductory, casual, or infrequent use of software.
>
> On the other hand, having to drill one's way through successive layers of menus and dialog boxes can be tedious and can make it difficult to reproduce a statistical analysis, perhaps with variations. Moreover, providing a GUI for a statistical system that includes hundreds (or even thousands) of commands, many incorporating extensive options, can produce a labyrinth (Fox, 2005, p. 2).

The opinion adopted in this monograph is that it is preferable to work with syntax using the standard R console or with script files, as discussed in Chapter 2. It is also worth noting that R can be run from Terminal on the Mac or the command window on Windows computers. These methods of running R are primarily intended to be used for batch use—several commands are issued for R to execute at a single time.

1.8   EXTENSION: MATRICES AND MATRIX OPERATIONS

Earlier in this chapter the vector was introduced as a unidimensional data structure used in R. As previously discussed, a vector is created using the c(), seq(), or rep() functions in R. Another common data structure is a *matrix*. In R, a matrix is a two-dimensional data structure, having both rows and columns. Matrices (the plural of matrix) are useful for displaying multivariate data and are also commonly used to mathematically express much of the computational aspects of applied statistics. In textbooks, matrix names are often typeset using bold, capital letters with the number of rows and columns listed below, separated by a "times" sign. For example, consider the following matrix:

$$\underset{3\times 4}{\mathbf{A}} = \begin{pmatrix} x_{11} & x_{12} & x_{13} & x_{14} \\ x_{21} & x_{22} & x_{23} & x_{24} \\ x_{31} & x_{32} & x_{33} & x_{34} \end{pmatrix}.$$

Each x_{ij} indicates an element, or value in the matrix, and the subscripts i and j indicate the element location—x is located in the ith row and jth column of matrix **A**. Since the number of rows and columns vary from matrix to matrix, the number of rows and columns, or *dimensions*, are listed along with the matrix name. Matrix **A** is a 3×4 matrix, having three rows and four columns.

A matrix is defined in R with the `matrix()` function. The function can take several arguments including `data=`, which is a user-provided vector of elements that will be converted to a matrix. The `nrow=` and `ncol=` arguments indicate the number of rows and columns, respectively, that the matrix will have. The `byrow=` argument indicates how the vector elements are to be filled into the matrix. When `byrow=FALSE`, the default value for the argument, the elements are filled into successive columns of the matrix starting with the first column. When `byrow=TRUE`, the elements are filled into successive rows of the matrix starting with the first row. Command Snippet 1.18 shows the syntax for converting a vector of data that includes the integer values 1 to 12 to a matrix with three rows and four columns.

Once a matrix has been defined, then its elements can be accessed selectively using an indexing system. The indexing system uses brackets separated by a comma, [,]. The space before the comma refers to the rows, and the space after the comma refers to columns. For example, to access (index) the value of the element in the fourth row and second column of matrix **B**, the syntax B[2,4] is used. It can be verified in Command Snippet 1.19 that this syntax will return the value of 8.

In addition to extracting a particular element, this type of indexing can also be used to change an element's value by assigning a new value to the indexed element. Command Snippet 1.19 also shows the syntax to both index the element in the second row and fourth column of **B** and change its value to 50. Entire rows or columns are indexed by a blank in either the rows or columns space. For example, B[2,] is used to index the entire second row of **B**, and B[,4] is used to index the entire fourth column. Additional examples are presented in Chapter 4.

1.8.1 Computation with Matrices

Addition of matrices can also be carried out as long as the matrices being added have the same dimensions. This is because adding two matrices **A** and **B** together requires adding their corresponding elements. This produces another matrix that has the same dimensions. For example,

$$
\begin{array}{c}
\mathbf{A} + \mathbf{B} \\
{\scriptstyle 3\times4} \quad {\scriptstyle 3\times4}
\end{array}
=
\begin{pmatrix}
1 & 4 & 7 & 10 \\
2 & 5 & 8 & 11 \\
3 & 6 & 9 & 12
\end{pmatrix}
+
\begin{pmatrix}
1 & 2 & 3 & 4 \\
5 & 6 & 7 & 50 \\
9 & 10 & 11 & 12
\end{pmatrix}
$$

$$
=
\begin{pmatrix}
2 & 6 & 10 & 14 \\
7 & 11 & 15 & 61 \\
12 & 16 & 20 & 24
\end{pmatrix}
$$

Command Snippet 1.18: Example of converting a vector of data that includes the integer values 1 to 12 to a matrix with three rows and four columns. For the first example, byrow=FALSE the elements are filled into successive columns of the matrix starting with the first column. In the second example, byrow=TRUE fills the elements into successive rows of the matrix starting with the first row.

```
## Vector of values 1 to 12
> X <- 1:12
> X
 [1]  1  2  3  4  5  6  7  8  9 10 11 12

## Put the values in a matrix by columns
> A <- matrix(data = X, nrow = 3, ncol = 4, byrow = FALSE)
> A
     [,1] [,2] [,3] [,4]
[1,]    1    4    7   10
[2,]    2    5    8   11
[3,]    3    6    9   12

## Put the values in a matrix by rows
> B <- matrix(data = X, nrow = 3, ncol = 4, byrow = TRUE)
> B
     [,1] [,2] [,3] [,4]
[1,]    1    2    3    4
[2,]    5    6    7    8
[3,]    9   10   11   12
```

R performs arithmetic computation on matrices using the addition (+) operator. Subtraction is carried out in a similar manner, except that one subtracts corresponding elements of the matrices. In R, the subtraction operator, $-$, is used to perform matrix subtraction. Command Snippet 1.20 shows the syntax for adding and subtracting two matrices in R.

The operation of scalar and matrix multiplication can also be carried out on matrices. A matrix can be multiplied by a single value, or *scalar*, by multiplying each of the elements in the matrix by that scalar. For example,

$$
\begin{array}{c}
3\,\mathbf{A} \\
{\scriptstyle 3\times4}
\end{array}
= 3
\begin{pmatrix}
1 & 4 & 7 & 10 \\
2 & 5 & 8 & 11 \\
3 & 6 & 9 & 12
\end{pmatrix}
=
\begin{pmatrix}
3 & 12 & 21 & 30 \\
6 & 15 & 24 & 33 \\
9 & 18 & 27 & 36
\end{pmatrix}
$$

To multiply a matrix by a scalar in R, we use the multiplication operator (∗). Command Snippet 1.21 shows the syntax for multiplying a matrix by a scalar in R.

Command Snippet 1.19: Syntax to index the element in the second row and fourth column of **B**. We also change the element's value to 50 using assignment.

```
## Index the element in the second row and fourth column
> B[2, 4]
[1] 8

## Change the element to 50
> B[2, 4] <- 50

## Verify it was changed
> B
     [,1] [,2] [,3] [,4]
[1,]    1    2    3    4
[2,]    5    6    7   50
[3,]    9   10   11   12
```

Command Snippet 1.20: Example of adding and subtracting two matrices.

```
## Matrix addition
> A + B
     [,1] [,2] [,3] [,4]
[1,]    2    6   10   14
[2,]    7   11   15   61
[3,]   12   16   20   24

## Matrix subtraction
> A - B
     [,1] [,2] [,3] [,4]
[1,]    0    2    4    6
[2,]   -3   -1    1  -39
[3,]   -6   -4   -2    0
```

Command Snippet 1.21: Example of multiplying a matrix by a scalar.

```
> 3 * A
     [,1] [,2] [,3] [,4]
[1,]    3   12   21   30
[2,]    6   15   24   33
[3,]    9   18   27   36
```

Multiplication of two matrices is more complicated requiring "row-column multiplication." Corresponding elements in the ith row of **A** are multiplied by the jth column of **B** and summed to produce the element in the ith row and jth column of the product matrix. This sounds complicated, but an example should help make things more clear. Consider the product of the matrices **X** and **Y** below:

$$\underset{2\times2}{\mathbf{X}} = \begin{pmatrix} 0 & 10 \\ 6 & 5 \end{pmatrix} \quad \underset{2\times2}{\mathbf{Y}} = \begin{pmatrix} 3 & 2 \\ 7 & 9 \end{pmatrix}$$

The element in the first row and first column of the product matrix would be found by computing $0 \times 3 + 10 \times 7 = 70$. Applying this rule for the remaining elements gives

$$\underset{2\times2}{\mathbf{X}}\underset{2\times2}{\mathbf{Y}} = \begin{pmatrix} 0 \times 3 + 10 \times 7 & 0 \times 2 + 10 \times 9 \\ 6 \times 3 + 5 \times 7 & 6 \times 2 + 5 \times 9 \end{pmatrix} = \begin{pmatrix} 70 & 90 \\ 53 & 57 \end{pmatrix}.$$

Multiplying two matrices together requires that they be *conformable*. This means the number of columns of the first matrix must be equal to the number of rows of the second matrix. To multiple X and Y, the syntax is X %*% Y. Command Snippet 1.22 shows the syntax for multiplying two matrices.

Command Snippet 1.22: Example of multiplying two matrices using the matrix multiplication operator.

```
## Create matrix A using the data in vector X
> X <- c(0, 10, 6, 5)
> A <- matrix(data = X, nrow = 2, ncol = 2, byrow = TRUE)
> A
     [,1] [,2]
[1,]    0   10
[2,]    6    5

## Create matrix B using the data in vector Y
> Y <- c(3, 2, 7, 9)
> B <- matrix(data= Y, nrow = 2, ncol = 2, byrow = TRUE)
> B
     [,1] [,2]
[1,]    3    2
[2,]    7    9

## Matrix multiplication
> A %*% B
     [,1] [,2]
[1,]   70   90
[2,]   53   57
```

1.9 FURTHER READING

Becker (1994) presents an interesting account of the history of the S language, which was the foundation for the R language. Venables and Ripley (2002) provides a classical, albeit terse, reference book to the S and R languages. For a fairly complete compendium to the R environment, including several examples, see Crawley (2007). There are also several free online resources. Two of note include the R Wiki (`http://wiki.r-project.org/rwiki/doku.php`)—a site "dedicated to the collaborative writing of R documentation," which provides more useful examples and generally has lengthier explanations than the official R documentation—and the Contributed Documentation section of the CRAN website.

PROBLEMS

1.1   An overweight male on a diet plan weighed himself once a month throughout the year. Suppose these weights from January to December were:

> 230, 226, 220, 217, 215, 215, 210, 204, 201, 199, 195, 190

Enter these data into a variable called *weight*. Use built-in R functions to compute the answers to the following questions.

    a) Create a variable called *loss* using the `diff()` function to produce a vector of the differences between each successive month's weight.

    b) What is the smallest amount of monthly weight loss?

    c) What is the largest?

    d) How many months was the amount of weight loss greater than 4 pounds?

    e) What is this person's average monthly weight loss?

    f) Use the `sum()` function to find the total weight loss over the year.

1.2   Use R to create the vectors **u** and **v**, where

$$\mathbf{u} = \begin{pmatrix} 1 \\ 2 \\ 3 \\ 4 \\ 5 \end{pmatrix} \quad \mathbf{v} = \begin{pmatrix} 1 \\ 4 \\ 7 \\ 10 \\ 13 \end{pmatrix}.$$

Use these vectors to perform the following computations in R.

    a) `u + 3`

    b) `v - 1`

    c) `length(u)`

    d) `length(v)`

    e) `sum(u > 3)`

    f) `sum(v[v > 5])`

    g) `prod(u < 2 | v > 7)`   read "|" as "or"

    h) `prod(u > 2 & v < 7)`   read "&" as "and"

    i) `v[5]`

    j) `v[u]`

    k) `v[v >= 5]`

1.3   Use R to create the vector **x**, where

$$\mathbf{x} = \begin{pmatrix} 1 \\ 4 \\ 5 \\ 8 \\ 3 \\ 2 \\ 9 \\ 6 \\ 6 \\ 1 \end{pmatrix}.$$

Use this vector to perform the following computations in R.

a) Use the sum() function to find the sum of the elements in x divided by 10.

b) Use the log() function to find the natural log for each of the elements in x.

c) Use chaining to compute the z-score for each element in x. (Try to complete this in one line of R syntax.) A z-score is computed using

$$z = \frac{x_i - \bar{x}}{s_x},$$

where x_i is an individual element in x, $\bar{x}$ is the average or mean of all of the elements in x, and s_x is the standard deviation of all of the elements in x.

d) Use the sort() function to sort the elements in x in ascending order.

e) Use the sort() function to sort the elements in x in descending order.

1.4   Use R to create the matrices **X** and **Y**, where

$$\underset{2\times 2}{\mathbf{X}} = \begin{pmatrix} 3 & 0 \\ 2 & 3 \end{pmatrix} \qquad \underset{2\times 2}{\mathbf{Y}} = \begin{pmatrix} 9 & 3 \\ 5 & 4 \end{pmatrix}$$

Use these matrices to perform the following computations in R.

a) Use the matrix multiplication operator to find the product **XY**.

b) Use the multiplication operator (∗) to compute **X**∗**Y**. Describe the computation that the multiplication operator performs on matrices.

CHAPTER 2

DATA REPRESENTATION
AND PREPARATION

The word data is the plural of the Latin datum, meaning a given, or that which we take for granted and use as the basis of our calculations ... We ordinarily think of data as derived from measurements from a machine, survey, census, test, rating, or questionnaire—most frequently numerical. In a more general sense, however, data are symbolic representations of observations or thoughts about the world.

—L. Wilkinson (2006)

Data stored outside of R—external to R—comes in many formats. It can be stored in a database, or it can be stored electronically, within a web page itself. For example, Zillow (http://www.zillow.com) keeps data on many variables related to real estate prices, which can be accessed over the Internet (see Example 2.1).

Data can also take on a very different look than just numbers, for example, medical researchers studying genetic information in the form of gene sequences (e.g., CAG, GGA, etc.). Other researchers might use data that takes the form of multimedia files— images (MRI scans, see Figure 2.1), audio files (speech recognition or music classification), or video files (facial recognition software).

Example 2.1 and Figure 2.1 show how data in some of these formats may look. Many devices from smart phones, to TiVos, to online accounts, such as Netflix and

Comparing Groups: Randomization and Bootstrap Methods Using R
First Edition. By Andrew S. Zieffler, Jeffrey R. Harring, & Jeffrey D. Long
Copyright © 2011 John Wiley & Sons, Inc.

Example 2.1. Real estate data stored in the XML code within the Zillow webpage

```
<?xml version="1.0" encoding="utf-8"?>
<SearchResults:searchresults xsi:schemaLocation="http://
www.zillow.com/static/xsd/SearchResults.xsd /vstatic/
71a179109333d30cfb3b2de866d9add9/static/xsd/SearchResults.xsd"
xmlns:xsi="http://www.w3.org/2001/XMLSchema-instance"
xmlns:SearchResults="http://www.zillow.com/static/xsd/
SearchResults.xsd">
    <request>
        <address>123 Bob's Way</address>
        <citystatezip>Berkeley, CA, 94217</citystatezip>
    </request>

    <message>
        <text>Request successfully processed</text>
        <code>0</code>

    </message>

    <response>
  <results>

  <result>
    <zpid>1111111</zpid> <links>
```

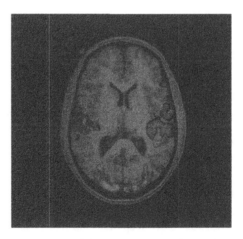

Figure 2.1: MRI scan stored as an image file.

Google, are storing data. While the data may not seem useful to educational and behavioral researchers, data are being used to remember who telephoned whom or to make predictions about which movie or TV show one may enjoy. The types of data being stored, the analyses being performed, and the questions that have begun to emerge about privacy and data ownership are something that educational and behavioral researchers may do well to begin considering.

2.1   TABULAR DATA

Although educational and behavioral researchers use many data structures in their research, most typically, they arrange data in a tabular fashion. In a tabular structure, each row represents a particular participant or case, and each column represents a variable. Here is an example:

ID	Achieve	ImmYear	ImmAge	English	Mex
1	68.90	68.20	5.00	1	0
2	70.20	80.40	6.30	1	0
3	65.00	75.70	1.80	1	1
4	78.00	73.80	2.30	1	1
5	81.90	62.10	7.10	1	0
6	83.20	82.00	8.10	1	0

In this data set, each row represents a single participant, a student in this case, and each column is an attribute of that student having been measured and recorded. It is common to use a spreadsheet program to enter these type of data.

2.1.1   External Formats for Storing Tabular Data

Popular spreadsheet programs for entering data in a tabular structure include *Excel* or *Calc*.[1] *Calc* is part of the *OpenOffice* suite of programs very similar to Microsoft Office. The OpenOffice website is http://www.openoffice.org/ where the entire suite can be freely downloaded. Another free spreadsheet program for Windows users is *Gnumeric* which is part of the GNOME project and is available at http://www.gnome.org/projects/gnumeric/downloads.shtml. Many popular data analysis programs (e.g., Minitab, Stata, SPSS) also include a spreadsheet-like editor for data entry. All of these programs have a *package-specific* format for storing tabular data. For example, *Excel* uses the **.xls* or **.xlsx* format and *SPSS* uses the **.sav* format.

Spreadsheet programs such as *Excel* and *Calc* are useful for entering and editing tabular data. Their native file formats, **.xls*, **.xlsx*, and **.odf*, however, typically go far beyond a simple tabular organization. They include separate sheets, or pivot

[1]*Excel* is a commercial program from Microsoft.

tables, and so there is no simple way to read these formats.[2] A better option is to use a *package-independent* format for storing tabular data. Two popular package-independent formats for tabular data storage are *.csv and *.txt.

Both *.csv and *.dat are *delimiter-separated value* formats. Delimiter-separated value formats store tabular data by separating or delimiting each value in the data in a specific way. Often the separator is a character, such as a colon or comma, or sometimes it is simply white space (e.g., a single space or tabbed space). A new case is usually delimited by a line break. Figure 2.2 shows data stored in a comma-separated value (*.csv) format.

```
ID,Achieve,ImmYear,ImmAge,English,Mex
1,68.9,68.2,5,1,0
2,70.2,80.4,6.3,1,0
3,65,75.7,1.8,1,1
4,78,73.8,2.3,1,1
5,81.9,62.1,7.1,1,0
6,83.2,82,8.1,1,0
7,76.1,72.6,7.7,0,1
8,46.8,64.2,8.4,1,0
9,63.1,72.3,5.8,0,1
10,64.4,67.3,1.5,1,0
```

Figure 2.2: Latino education data opened in *Text Editor* on the Mac. The data are delimited by commas and each case is delimited by a line break

2.2 DATA ENTRY

During the data collection and data entry processes, it is important to consider the structure of the data before reading it into R. Here are some guidelines that can be used to prepare or edit data while working in a spreadsheet.

- All of the data should be entered in a single sheet in one file.

- Enter variable names in the first row of the spreadsheet.

- Variable names should be reflective/descriptive of the measure being represented.

- Use variable names that begin with a letter.

[2]There are functions available to read data from both formats into R—read.xls() from the **RODBC** library and read.odf() from the **ROpenOffice** library available at http://www.omegahat.org/ROpenOffice/.

- Variable names cannot contain spaces.

- Always include an ID variable.

- If a variable has multiple groups, include additional variables that indicate group membership.

- For missing values, leave the cell blank.

2.2.1 Data Codebooks

When data are entered or prepared, a *codebook* should also be developed. A codebook is a technical description of the data collected for a particular purpose. It describes how the data are arranged in the computer file, what the various values indicate, and any special instructions on proper use of the data. At a minimum, codebooks should include the following:

- Description of the study (e.g., how the study was performed, population studied, how the sample was selected, etc.)

- Source (e.g., who were the researchers that collected the data)

- Technical information about the data set (e.g., number of observations, etc.)

- Variable names (e.g., what does `MultAcc` mean?)

- Variable values (e.g., 1 = male and 2 = female)

- Missing data codes (e.g., -999 for nonresponse, -888 for inability to locate study participant)

- Measurement information for each variable (e.g., units and ranges)

Consider the codebook for the data stored in *LatinoEd.csv*. These data are used in Chapter 7 to explore potential educational achievement level differences between Mexican and non-Mexican Latino immigrants. The codebook for these data is shown below.

Other example codebooks are available at the following locations: `http://global.sbs.ohio-state.edu/docs/Codebook-12-12-05.pdf` and `http://nces.ed.gov/nationsreportcard/tdw/database/codebooks.asp`

2.3 READING DELIMITED DATA INTO R

When using R, it is convenient to store the data in a file external to R, and then analyze it after "reading it into" R. The data are not analyzed in the same "place" as they reside, as with a spreadsheet-style program. This way, the original data never need to be altered for the purposes of analysis. In order to read delimited data into R, the following characteristics need to be identified:

Data Codebook

Overview: The data come from Stamps and Bohon (2006) who were predicting variation in educational achievement of Latinos. The 150 Latinos included in the sample are naturalized U.S. citizens living in Los Angeles and were drawn from data collected during the 2000 U.S. Census. Variables included in the data set are the following:

- `ID`: ID number.

- `Achieve`: Educational achievement level. This is a scale of educational achievement, ranging from 1 to 100, in which higher values indicate higher levels of educational achievement.

- `ImmYear`: Year in which the immigrant arrived in the United States. (This indicates the year—to the nearest tenth—in which the immigrant arrived. To get the full year, 1900 must be added to each value. For example, 81.5 is half way through 1981.)

- `ImmAge`: Immigrant's age at time of immigration (age to nearest tenth of a year).

- `English`: Is the individual fluent in English? (0=No; 1=Yes)

- `Mex`: Did the individual immigrate from Mexico? (0=No; 1=Yes.)

Reference: Stamps & Bohon (2006). Educational attainment in new and established Latino metropolitan destinations, *Social Science Quarterly, 87*(5), 1225–1240.

- Where the data are located on the computer

- How the values are delimited

- Whether or not variable names have been included in the first row of the data file

2.3.1 Identifying the Location of a File

External data files can reside on the computer or the web. In order to read the data into R, the location of a file must be supplied, either as a file location in a directory tree or as an Internet address. Files are typically stored on a computer using a type of folder structure (see Figure 2.3). By double-clicking through this structure, files can be located and opened. R does not locate files using this type of navigation. Instead, R needs the literal location of the file specified in terms of the *path* and *filename*.

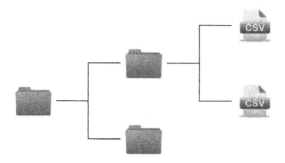

Figure 2.3: Folder and file hierarchy on a computer system.

The term *path* may be novel to some readers because the graphical user interface of the most popular operating systems have been designed to give the look and feel of "picking" a file. The path is simply the navigation of the folder or *directory* hierarchy provided as a character string.

The filename is a means of storing metadata about a particular file stored on the computer via a character string (e.g., *mydata.csv*). The metadata that is human readable often includes both the *basename* of the file itself and the *extension*. The basename is the primary filename (e.g., *mydata*), while the extension (e.g., **.csv*)— which is optional on some operating systems—indicates how the data stored in the file were encoded.

To find the path and filename of a data file, the function `file.choose()` is used with no arguments (see Command Snippet 2.1). Execution of the function opens a window showing the file structure, which can be used to navigate to and select the file of interest. Once a file is selected by either double-clicking its name or clicking OK in the window, the path name, filename, and file extension is ported to the R console.

Command Snippet 2.1: Finding the path name for the *VLSSAge.csv* file.

```
> file.choose()
[1] "/Documents/Data/LatinoEd.csv"
```

The syntax in Command Snippet 2.1 is for an Apple computer. The syntax for a Windows computer uses a backslash (\) rather than the forward slash (/) to separate directories.[3] Command Snippet 2.2 shows the use of `file.choose()` and the output based on the same directory hierarchy, but using Windows.

---

[3]Windows OS, which followed from MS-DOS 2.0 introduced the backslash as the directory separator in the early 1980s. Although Unix, which the MacOS is based on, had previously introduced the forward slash as the directory separator sometime around 1970, MS-DOS 1.0 was already using the forward slash to introduce command-line options so they used a different character for their directory separator.

Command Snippet 2.2: Finding the path name for the *VLSSAge.csv* file on Windows.

```
> file.choose()
[1] "C:\\Documents\\Data\\LatinoEd.csv"
```

One thing to note about the path name used by R on Windows computers is that the directory separator uses *two* backslashes. This is in contrast to the single backslash that is standard for the Windows operating system. The backslash is a character that has special meaning when used in a character string. It is called an escape character. Thus, in order to let R know the intention is to specify a file location rather than an escape sequence, the double backslashes must be used.

2.3.2   Examining the Data in a Text Editor

If the data to be analyzed was not entered by the same person conducting the analysis, then aspects of the data file might be unclear. Depending on the program used to encode the data, the filename extension can be a useful first step in helping to identify the delimiter used to separate the values. Often the extension **.csv* is used for comma delimiting and **.txt* is used for space delimiting. It is sometimes worthwhile to also examine the file via a text editor to visually assess which delimiter is used.[4] Besides visually identifying the value delimiter, opening the data in a text editor also allows the determination of whether the variable names are included in the first row of the file.

The Latino education data has a file suffix of **.csv*. A portion of the contents is shown in Figure 2.3. The figure shows the Latino education data opened in *Text Editor* on the Mac. Figure 2.3 indicates each datum is separated by a comma, and the variable names are included in the first row. This information is important for proper reading of the data.

2.3.3   Reading Delimited Separated Data: An Example

To read in delimited data, the read.table() function is used. This function requires the argument file=. The argument takes the appropriate path and filename. The path and filenames are character strings. As such, they should be enclosed in quotation marks. For example,

file="/Documents/Data/LatinoEd.csv"

Two other arguments required in many cases are header= and sep=. The header= argument takes a logical value of either TRUE or FALSE depending on whether the variable names are included in the first row of the file. Since the Latino data includes variable names in the first row of the data, the argument will be header=TRUE.

---

[4]*TextEdit* on the Mac and *Notepad* on Windows are good choices for this activity.

The `sep=` argument provides a character string that identifies the delimiter used to separate the values. Since the values in the Latino data are separated by commas, the argument will be `sep=","`. The help menu for `read.table()` reveals that the default values for these arguments are `header=FALSE`—there are not variable names in the first row of the data—and `sep=" "`—the value delimiter is one or more "white" spaces or blank spaces.

Lastly, every time data is read into R, it is a good idea to assign the data to an object. This makes it easier to manipulate and analyze the data set. In this example, suppose the data are assigned to the object `latino`. The full command to read in the data and assign it is given in Command Snippet 2.3.

Command Snippet 2.3: Using the `read.csv()` function to read in the data stored in the *LatinoEd.csv* file.

```
> latino <- read.table(file = "Documents/Data/LatinoEd.csv",
      header = TRUE, sep = ",")
```

It can be said that R operates on a "no news is good news" convention. After executing the syntax in Command Snippet 2.3, it appears as if nothing happened! No spreadsheet opens up and there is no message of confirmation the data were correctly read. Error messages and possible warning messages are typically issued if something goes wrong, but nothing is output when things go right. In this case, the *LatinoEd.csv* file has been successfully read into R and the program is waiting for the next command.

2.4 DATA STRUCTURE: DATA FRAMES

When delimited, tabular data are read into R and assigned to an object, that object is stored as a *data frame*. The data frame is another fundamental data structure used in R. A data frame is a rectangular array, in which all of the columns have the same length—similar to a matrix. Unlike matrices, however, data frames can have columns of different types (e.g., numeric, factor, character, etc.). Because of this, data frames are a better structure in which to store data that consist of both quantitative and categorical variables. There are several R functions that have specific methods and functionality for working with data frames. Some of these functions are described in this section.

2.4.1 Examining the Data Read into R

Once data are read into R, manipulation and analysis is performed with the created data frame (e.g., the `latino` object) rather than the original data file. Prior to any analysis performed in R, the newly created data frame should be checked to help ensure the data were correctly read. Examining the data frame also reveals how R is treating variables. For example, are numerically coded factors being treated as numbers or factors? Examination of the data frame is also helpful for beginning the process of data cleaning (e.g., finding errors in transcription or identifying problematic observations).

By examining the first few rows of the data frame and the last few rows of the data frame, it can often be determined whether the data were properly read into R. This is done by calling the head() and the tail() function on the data frame object, in this case latino (see Command Snippet 2.4).[5]

Command Snippet 2.4: Examining the latino data frame.

```
## Examine the first part of the data frame
> head(latino)
  ID Achieve ImmYear ImmAge English Mex
1  1    59.2    77.7    9.6       1   1
2  2    63.7    65.8    1.1       1   1
3  3    62.4    63.6    6.1       0   1
4  4    46.8    55.3    2.1       1   1
5  5    67.6    73.1    2.3       1   1
6  6    63.1    75.7    8.4       1   0

## Examine the last part of the data frame
> tail(latino)
     ID Achieve ImmYear ImmAge English Mex
145 145    76.1    79.4    9.9       1   1
146 146    70.2    62.8    5.5       1   0
147 147    41.6    70.7    0.8       0   1
148 148    70.2    82.0   10.4       1   1
149 149    63.1    56.6    1.2       1   0
150 150    81.9    66.0    0.7       1   0
```

Whenever data frames are printed in R (e.g., using the head() function), the row names are printed along with each of the variables. The row names are printed in the leftmost column. In this example, the row names are redundant to the information in the ID variable. When the original data set has a variable that can be used for identification (e.g., ID, names, etc.), it is sometimes convenient to use this variable for the row names. This is accomplished by using the row.names= argument in the read.table() function. This argument takes a character string indicating the name of the variable to be used as the row names (e.g., row.names="ID"), or the column number, which contains the values to be used as the row names (e.g., row.names=1 can be used). Command Snippet 2.5 shows the use of the row.names= argument in the read.table() function to use the values in the ID variable as the row names of the latino data frame.

The output in Command Snippet 2.5 shows that when the latino data frame is printed now, there is no ID variable. This is because the ID variable is now being used

---

[5]Both the head() and tail() functions can be provided an optional argument, n=, which takes a numeric value to indicate how many cases will be printed. For example, head(latino, n = 20) will print the first 20 cases rather than the first 6 cases.

Command Snippet 2.5: Examining the latino data frame after using
row.names="ID".

```
## Read in the Latino data
> latino <- read.table(file = "Documents/Data/LatinoEd.csv",
    header = TRUE, sep = ",", row.names = "ID")

## Examine the first part of the data frame
> head(latino)
  Achieve ImmYear ImmAge English Mex
1   59.2    77.7    9.6      1    1
2   63.7    65.8    1.1      1    1
3   62.4    63.6    6.1      0    1
4   46.8    55.3    2.1      1    1
5   67.6    73.1    2.3      1    1
6   63.1    75.7    8.4      1    0

## Examine the last part of the data frame
> tail(latino)
    Achieve ImmYear ImmAge English Mex
145   76.1    79.4    9.9      1    1
146   70.2    62.8    5.5      1    0
147   41.6    70.7    0.8      0    1
148   70.2    82.0   10.4      1    1
149   63.1    56.6    1.2      1    0
150   81.9    66.0    0.7      1    0
```

as the row names for the data frame object. The ID variable has not disappeared from
the original data, it is just no longer needed as a variable within R itself. If the function
row.names() is used on the latino data frame (e.g., row.names(latino)), R will
print the vector of row names, which for all intents and purposes is the ID variable.
The ID variable can be re-created as a new column in the data frame by using the
syntax latino$ID <- rownames(X = latino).

After looking at the first and last few observations, the internal structure of the
data frame should be examined by use of the str() function. Recall that a data
frame can contain columns that are of different types, or *classes*. The str() function
will output the class for each of the columns (variables) in the data frame. Command
Snippet 2.6 shows the results of calling the str() function on the latino data frame.

The result from the str() function provides a lot of information. It provides the
data structure of the object being called—in this case a data frame—as well as the
number of rows (150 observations) and columns (5 variables). The last part of the
output lists each of the column names in the data frame, the class of each of these
variables, and the values for the first several cases. For example, R is treating these
variables as either numeric (num) or integer (int). Since certain functions in R can
only be used on variables of certain classes, it is very important to determine the class
for each variable.

Lastly, it is good to examine the data for observations that might have been mis-
keyed or that could be unusual. The summary() function is a useful first step in this

Command Snippet 2.6: Examining the internal structure of the `latino` data frame.

```
> str(latino)
'data.frame': 150 obs. of  5 variables:
 $ Achieve: num   59.2 63.7 62.4 46.8 67.6 63.1 63.7 63.1 67.6
      30.6 ...
 $ ImmYear: num   77.7 65.8 63.6 55.3 73.1 75.7 72 69.9 63.9
      77.5 ...
 $ ImmAge : num   9.6 1.1 6.1 2.1 2.3 8.4 4.9 5.2 12.7 9.2 ...
 $ English: int   1 1 0 1 1 1 0 1 1 0 ...
 $ Mex    : int   1 1 1 1 1 0 1 1 1 1 ...
```

exploratory process. Using the `summary()` function on a data frame object elicits summary information for each of the variables within the data frame, including the values for the minimum, maximum, first and third quartiles, and the mean.[6] Command Snippet 2.7 shows the results of using the `summary()` function on the `latino` data frame.

Command Snippet 2.7: Examining each variable in the `latino` data frame using the `summary()` function.

```
> summary(latino)
    Achieve          ImmYear           ImmAge
 Min.   :21.50   Min.   :15.30   Min.   : 0.000
 1st Qu.:52.00   1st Qu.:65.83   1st Qu.: 3.675
 Median :62.40   Median :72.25   Median : 6.450
 Mean   :59.94   Mean   :69.98   Mean   : 6.613
 3rd Qu.:69.42   3rd Qu.:77.80   3rd Qu.: 9.550
 Max.   :96.20   Max.   :83.40   Max.   :12.900

    English            Mex
 Min.   :0.0000   Min.   :0.0000
 1st Qu.:0.2500   1st Qu.:1.0000
 Median :1.0000   Median :1.0000
 Mean   :0.7467   Mean   :0.7733
 3rd Qu.:1.0000   3rd Qu.:1.0000
 Max.   :1.0000   Max.   :1.0000
```

A cursory examination of the summary information suggests that the values for each of the variables seem reasonable based on the codebook. For example, the minimum and maximum values fall within expected limits based on the nature of each variable. It appears the data are ready for analysis.

---

[6]The `summary()` function deals with classes of objects differently. For example, calling `summary()` on a data frame produces different output than when it is called on an object that is of the `lm` class.

2.5   RECORDING SYNTAX USING SCRIPT FILES

As an R session progresses, it is desirable to record and save syntax so the commands can be accessed at a future point in time. One of the principal reasons for doing so is that the session can be exactly reproduced at some later time. Reproducibility is a major advantage of syntax-based statistical programs such as R. One way to record commands is to save them in a *script file*. A script file not only allows R syntax to be recorded, it also provides the foundation for documentation, debugging, revision, and validation. A script file is often created using a script *editor*. Any editor is permissible with R, but some are better than others.[7]

R comes with a simple script editor, which can be accessed using the menu system within R. Select the File menu and then New script (New Document on a Mac), which will open another window that is blank. In this blank window, commands can be saved and executed or copied and pasted within or out of R. It is required that separate commands be written on separate lines or separated by a semicolon (;) if written on the same line. Furthermore, the script file must contain only valid commands—no prompts or output are allowed unless contained in a commend that is not read by R. Comments are preceded by at least one hash or pound sign (#). Anything following a pound sign to the end of the current line will not be read by R.

Script File with Comments

```
## Read the Latino data into R
latino <- read.table(file = "Documents/Data/LatinoEd.csv",
    header = TRUE, sep = ",", row.names = "ID")

## Examine the Latino data head(latino)
tail(latino)
str(latino)
summary(latino)

## Graph the educational achievement scores
boxplot(latino$Achieve)
plot(density(latino$Achieve))
```

The main advantage of the script file is to save a record of the analyses, which can be replicated in future R sessions. Another advantage in recording commands in a script file is that the entire file of syntax or chunks of the syntax can be submitted to R for processing directly. To submit the script for processing, highlight the commands

[7]There are a number of R script editors available free of charge. A list is provided at: http://www.sciviews.org/_gui/projects/Editors.html. Many of these script editors know about R syntax and help in creating permissible syntax, such as matching parentheses, etc.

to be executed, and from the menu select Edit and then Run line or selection (Execute on a Mac). The commands will appear in the console window and will be processed one by one. Alternatively, key strokes can be used for the submission. On a Windows machine, highlighted syntax is submitted by using Ctrl-r. On a Mac highlighted syntax is submitted by using Cmd-Return.

It is also a good idea to use *comments* to document the process in the script file. Comments are human-readable, explanatory text that is inserted into the script file along with the R commands. To insert a comment in the script file, start the line with the hash or pound sign (#). This informs R that everything after # on the current line is to be ignored and not executed.

2.5.1 Documentation File

In many analysis projects, it is useful to have a *documentation file* in addition to the script file. A documentation file is a word-processing file containing information (commentary, summaries, tables, figures) about the work being performed or the results from that work. The information is usually intended to be read by a human.[8] Ideally, a documentation file should give enough information about the related scripts so that it is possible to determine the connection between the two, and even to revise the documentation when the script is revised.[9]

2.6 SIMPLE GRAPHING IN R

There are many graphing packages within R. These include the basic **graphics** package, the **lattice** package, the **gplots** package, and the **ggplot2** package, just to name a few. Each package has its strengths and weaknesses. This monograph will focus on using the first as it is automatically installed with the base package.

Many of these packages implement some of the same plots, albeit in a different manner. For example, to plot a kernel density estimate, the **graphics** package uses the plot() function on the density() function, whereas the **lattice** package uses the function densityplot(). Similarly, a box-and-whiskers plot can be called by the function boxplot() in the **graphics** package, and bwplot() in the **lattice** library. There are, of course, many differences in the usage of these functions in terms of arguments provided, output produced, etc. Throughout the remainder of this monograph, the plotting capabilities in the **graphics** package will be explored in detail. For now, focus is on some basic ideas to continue with the familiarization to R.

One basic method for summarizing the distribution of a variable is the *box-and-whiskers plot* or *boxplot* for short. To create a simple box-and-whiskers plot of the

---

[8]Script files can also be read by humans, and sometimes they are a proxy for documentation files. But they are not suitable for presenting results.

[9]Scripts can also be integrated into documentation files. This is an advanced topic and one we will not cover in this monograph. The interested reader is referred to the reference on Sweave in the list of further reading.

achievement scores provided in the Latino data set, the `boxplot()` function from the **graphics** package is used. This package is automatically loaded when a R session is initiated. The `boxplot()` function requires the argument x= giving the name of the variable to be plotted. Command Snippet 2.8 shows what would seem to be logical syntax for plotting the achievement scores.

Command Snippet 2.8: Creating a box-and-whiskers plot of the educational achievement scores in the Latino data set.

```
> boxplot(Achieve)
Error in boxplot(Achieve) : object 'Achieve' not found
```

Calling the `boxplot()` function on the variable `Achieve` produces an error. The problem is that R cannot find the object named `Achieve`. The reason is that `Achieve` is in the data frame object `latino`, and this must be referenced in order to access `Achieve`. To inform R that it needs to look "inside" the object `latino` to find `Achieve`, the $ operator is used, as in

`latino$Achieve`

Command Snippet 2.9 provides the correct syntax for creating a box-and-whiskers plot of the achievement scores.

Another basic type of graph useful for examining the distribution of a variable is the *density plot*. The second line of Command Snippet 2.8 provides the syntax for creating a density plot of the achievement scores using the **graphics** package. (Density plots are discussed in more detail in Chapter 3.) Figures 2.4 and 2.5 show the resulting plots from these commands.

Command Snippet 2.9: Creating a box-and-whiskers and density plot of the educational achievement scores.

```
## Box-and-whiskers plot
> boxplot(latino$Achieve)

## Density plot
> plot(density(latino$Achieve))
```

2.6.1 Saving Graphics to Insert into a Word-Processing File

Once graphs are created, it is desirable to include them in the word-processing document that describes the analysis. This document might be, for example, the draft of a professional journal article. There are several ways to insert a created plot into a word-processing document. One of the easiest is to *copy* the picture in R and *paste* it into the documentation file. On a Mac, after making sure that the R graphics Quartz

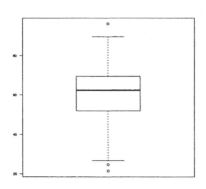

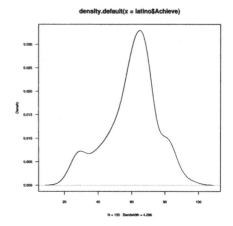

density.default(x = latino$Achieve)

Figure 2.4: Box-and-whiskers plot of the distribution of educational achievement scores for 150 Latino immigrants.

Figure 2.5: Density plot of the distribution of educational achievement scores for 150 Latino immigrants.

window is the active window (clicking on it brings it to the front), choose Copy from the Edit menu, and then paste it from within the documentation file. If Windows is used, right-click on the graphics window directly and choose Copy, and again paste it into the documentation file. This technique is fine for drafts and homework, but for publishable quality plots, this technique might produce a sub-par quality graphic. [10]

The problem is that the copy function does not choose a high-quality image format. The current default on most Macs is *.png*, whereas for most PCs it is *.jpeg*. Both of these formats are good choices for the copy function as their compression scheme provides a nice balance of moderate image quality and small file space. But, they do not provide the level of image quality that most journals would require in their printing process. This is because both of these image formats are *bitmapped* images.

Bitmapped images are essentially a map of bits or pixels. [11] If the image in question is conceived of as a grid, then certain pixels, or squares, in that grid are filled in to create the image. The grid can be made finer to look better at higher resolutions, but when zooming in or out on the image, the quality deteriorates. The reason is bitmapped images are dependent on the resolution at which they were created. For images to appear on the Web, bitmapping works very well, since most computer screens have a limited resolution. The size of the image is not restricted due to space considerations, so the author often has a lot of control over how the image will appear.

This dependency, however, makes bitmapped images a poor choice for graphics that are to appear in print, since many commercial printers can print at a very high resolution. What looks good on a computer screen may look terrible in print. A better

---

[10] Actually, using a golfing metaphor this produces an "above-par quality graphic."
[11] Bitmapped images are also sometimes referred to as Raster images.

option for images that are to appear in print is to use an image type that is *vector* formatted. Vector-formatted images save strokes (vectors) and, thus, will redraw correctly at a high resolution when the image is resized. Because the image is saving vectors, these image types can be slow to open and often take up more disk space than bitmapped images. Common vector formats for images include **.pdf*, **.eps*, and **.svg*.

R has several built-in functions to produce plots in various image formats. The plots in this monograph, for example, were produced as **.pdf* files. To save an image as a **.pdf* file, the `pdf()` function is used. The function takes the argument `file=`, which requires the path and filename for saving the image. Other useful arguments are `width=` and `height=` that can be used to specify the size of the created image in common units of measure (e.g., in, cm, pt, etc.).

The `pdf()` function opens a graphics driver in R, and then each subsequent R command writes to that PDF file. This continues to happen until the driver is turned off by calling the `dev.off()` function. The full set of commands to produce a density plot of the educational achievement scores and save it as a **.pdf* file are provided in Command Snippet 2.10. For a Windows-based computer, the path name must be specified with the double backslashes as previously described.

Command Snippet 2.10: Creating a density plot of the educational achievement scores in the Latino data set and saving it as a PDF file called *VLSS-Age.pdf* on the desktop on a Macintosh computer.

```
## Open the graphics driver
> pdf(file = "/Desktop/VLSS-Age.pdf")

## Draw the plot
> plot(density(latino$Achieve))

## Turn off the graphics driver
> dev.off()
```

2.7 EXTENSION: LOGICAL EXPRESSIONS AND GRAPHS FOR CATEGORICAL VARIABLES

Based on the Latino data set introduced earlier, suppose the goal is to examine the distribution of achievement scores for Latino immigrants who came to the United States between the ages of 4 and 7. Since the age at immigration is recorded in the variable `ImmAge`, this variable is used to filter the study participants. The achievement scores for the Latinos must meet the two simultaneous conditions of `ImmAge` $\geq$ 4 *and* `ImmAge` $\leq$ 7.

In R, *comparison operators* are used to evaluate a particular condition. For example, to evaluate whether an immigrant in the `latino` data frame was at least four years old when she or he immigrated to the United States, the following expression is used: `latino$ImmAge >= 4`. The comparison operators are shown in Table 2.1.

Table 2.1: Comparison Operators Used in R

Comparison Operator	Meaning
<	Less than
>	Greater than
<=	Less than or equal to
>=	Greater than or equal to
==	Equal to
!=	Not equal to

Expressions involving comparisons are called *logical expressions* because they return a logical value of TRUE or FALSE depending on the evaluation of the expression. For example, the logical expression, latino$ImmAge >= 4 is evaluated as TRUE for each Latino immigrant whose age at the time of immigration is greater than or equal to 4, and it is evaluated as FALSE otherwise. Command Snippet 2.11 shows the results when the expression latino$ImmAge >= 4 is evaluated for the 150 immigrants represented in the latino data frame.

Command Snippet 2.11: Evaluation of the logical expression latino$ImmAge >= 4 for the 150 immigrants represented in the latino data frame.

```
> latino$ImmAge >= 4
  [1]   TRUE FALSE   TRUE FALSE FALSE   TRUE   TRUE   TRUE   TRUE
 [10]   TRUE   TRUE   TRUE   TRUE FALSE   TRUE   TRUE   TRUE FALSE
 [19] FALSE FALSE   TRUE   TRUE   TRUE   TRUE   TRUE   TRUE   TRUE
 [28]   TRUE   TRUE   TRUE FALSE   TRUE   TRUE   TRUE FALSE   TRUE
 [37]   TRUE   TRUE   TRUE FALSE   TRUE   TRUE   TRUE   TRUE   TRUE
 [46]   TRUE   TRUE   TRUE FALSE   TRUE   TRUE   TRUE   TRUE   TRUE
 [55]   TRUE   TRUE   TRUE   TRUE FALSE FALSE   TRUE   TRUE   TRUE
 [64] FALSE FALSE   TRUE   TRUE   TRUE   TRUE   TRUE FALSE   TRUE
 [73] FALSE   TRUE   TRUE   TRUE   TRUE FALSE   TRUE   TRUE   TRUE
 [82]   TRUE FALSE FALSE FALSE FALSE FALSE   TRUE FALSE   TRUE
 [91]   TRUE   TRUE   TRUE   TRUE   TRUE   TRUE FALSE   TRUE   TRUE
[100]   TRUE   TRUE   TRUE FALSE   TRUE   TRUE FALSE   TRUE   TRUE
[109]   TRUE FALSE   TRUE   TRUE   TRUE   TRUE FALSE   TRUE   TRUE
[118]   TRUE   TRUE   TRUE   TRUE   TRUE   TRUE FALSE   TRUE   TRUE
[127] FALSE   TRUE   TRUE   TRUE FALSE   TRUE   TRUE FALSE   TRUE
[136]   TRUE   TRUE FALSE   TRUE   TRUE   TRUE FALSE   TRUE FALSE
[145]   TRUE   TRUE FALSE   TRUE FALSE FALSE
```

2.7.1 Logical Operators

Use of the logical expression latino$ImmAge >= 4 allows the determination of which of the 150 immigrants in the latino data frame have met the first condition, namely that ImmAge $\geq$ 4. However, the goal is to identify which immigrants meet this condition and the second condition, ImmAge $\leq$ 7. *Logical operators* are common computing tools to combine multiple logical expressions. A logical operator can be used along with the logical expression to determine the immigrants who were at least 4 years old but no older than 7 when they immigrated. Logical operators used in R are shown in Table 2.2.

Table 2.2: Logical Operators Used in R

Logical Operator	Meaning
&	And (intersection)
|	Or (Union)

The & operator is used to combine both the logical expression latino$ImmAge >= 4, and latino$ImmAge <= 7, to identify Latino immigrants who came to the United States between the ages of 4 and 7. Command Snippet 2.12 shows the syntax to evaluate these expressions.

Command Snippet 2.12: Use of the & operator to evaluate multiple logical expressions.

```
> latino$ImmAge >= 4 & latino$ImmAge <= 7
  [1] FALSE FALSE  TRUE FALSE FALSE FALSE  TRUE  TRUE FALSE
 [10] FALSE FALSE FALSE FALSE FALSE FALSE  TRUE FALSE FALSE
 [19] FALSE FALSE  TRUE  TRUE FALSE FALSE FALSE FALSE FALSE
 [28] FALSE  TRUE  TRUE FALSE  TRUE  TRUE FALSE FALSE FALSE
 [37]  TRUE FALSE  TRUE FALSE FALSE FALSE FALSE FALSE  TRUE
 [46] FALSE FALSE FALSE FALSE  TRUE  TRUE FALSE FALSE FALSE
 [55] FALSE FALSE FALSE  TRUE FALSE FALSE  TRUE FALSE FALSE
 [64] FALSE FALSE FALSE  TRUE FALSE  TRUE FALSE FALSE FALSE
 [73] FALSE  TRUE  TRUE FALSE FALSE FALSE  TRUE FALSE  TRUE
 [82] FALSE FALSE FALSE FALSE FALSE FALSE  TRUE FALSE  TRUE
 [91]  TRUE FALSE FALSE FALSE  TRUE FALSE FALSE FALSE  TRUE
[100]  TRUE  TRUE  TRUE FALSE FALSE  TRUE FALSE  TRUE FALSE
[109]  TRUE FALSE FALSE FALSE FALSE FALSE FALSE  TRUE FALSE
[118] FALSE FALSE  TRUE FALSE FALSE FALSE FALSE FALSE  TRUE
[127] FALSE  TRUE FALSE FALSE FALSE FALSE FALSE FALSE  TRUE
[136] FALSE  TRUE FALSE FALSE FALSE FALSE FALSE FALSE FALSE
[145] FALSE  TRUE FALSE FALSE FALSE FALSE
```

The & and | operators perform an element-wise or component-wise evaluation using the logical expressions. In other words, there is a logical value produced for each element in the latino$ImmAge vector. For example, the value for ImmAge for the first case in the latino data frame is 9.6. The logical value produced by the first component of the evaluation, latino$ImmAge >= 4, is TRUE. The logical value produced by the second component of the evaluation, latino$ImmAge <= 7, is FALSE. Since at least one of the components is FALSE, the intersection of these two logicals is FALSE.

There are two other logical operators, && and | |, that perform scalar evaluation, but this is not useful in the type of identification discussed here. Scalar evaluation is very useful for programming when it is desirable to check that certain conditions are met before another set of commands is executed. For example, if the goal is to be sure that *all* the elements in a vector are positive and less than 10, using x > 0 && x < 10 would ensure this. The second logical expression (x < 10) is only evaluated if the first logical expression evaluates to TRUE, and since this is a scalar evaluation, it only evaluates to TRUE if all of the elements are positive. Similarly, the second expression would also only evaluate to TRUE if all of the elements were less than 10. So, if any of the elements did not meet the conditions, the entire expression would produce a single logical of FALSE.

Logical operators are very useful for extracting a subset of the data for examination. These operators are commonly used within the subset() function. The function subset() can be used to extract all the Latino immigrants who came to the United States between the ages of 4 and 7. The first argument in the subset() function, x=, is the object to be subsetted, and the second argument, subset=, is the logical expression(s) defining the selection conditions. The output from the subset() function can be assigned to a new object, which will have the same structure as the original. Command Snippet 2.13 shows the use of the subset() function to identify the Latino immigrants who came to the United States between the ages of 4 and 7. All the data associated with these immigrants is assigned to a new object called latino.sub. The new object is examined using the summary() function. Note that the variable ImmAge has minimum and maximum values appropriate to the subsetting that occurred.

Any of the functions that have been introduced thus far for data examination, plotting, or analysis can be used in the latino.sub data frame. For example, a density plot of the distribution of achievement scores for these 41 immigrants who came to the United States between the ages of 4 and 7 years of age can be produced using the syntax plot(density((latino.sub$Achieve)). Figure 2.6 shows the resulting plot of the command.

2.7.2 Measurement Level and Analysis

Both a density plot and a box-and-whiskers plot were used to visualize the distribution of achievement scores for Latino immigrants. Both of these plots are appropriate graphics tools for summarizing data that are numerical—like achievement scores. These graphs, however, are inappropriate for summarizing data that are categorical (i.e., have categories). In general, appropriate methods for graphically and numerically summarizing data depend on the *measurement level* of the variable being summarized.

Paraphrasing N. R. Campbell, Stevens (1946, p. 677) wrote that "measurement, in the broadest sense, is defined as the assignment of numerals to objects or events according to rules." Building on this definition, he pointed out that different rules

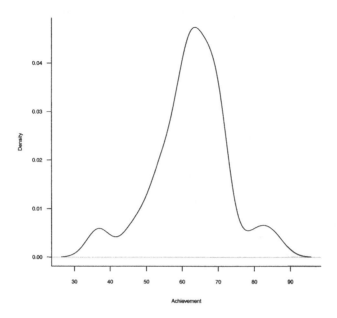

Figure 2.6: Density plot of the distribution of educational achievement scores for 41 Latinos who immigrated to the United States between the ages of 4 and 7.

for assigning numbers lead to varying mathematical properties and, subsequently, different statistical operations that can be employed. The classification that Stevens proposes includes four distinct measurement levels that increase in their sophistication: nominal, ordinal, interval, and ratio.

We now briefly explain the four classifications, but the more interested student should read the Stevens article or a modern treatment (e.g., Long, Feng, & Cliff, 2003). The *nominal* level of measurement is used for variables in which the assignment of numerals only represents labels or categories. Examples of nominal level variables are race, sex, or political preference. The *Ordinal* level of measurement is attributed to variables that can be rank ordered. In the educational and behavioral sciences, education level and score on a Likert item or scale (e.g., 1 = low, 2 = medium, 3 = high) are examples of ordinal data. The third level of measurement is the *interval* level of measurement. The numerals representing interval scales not only have a rank ordering to them, but each successive numeral represents the same "amount" of whatever is being measured. An example is calendar time where a change of 10 days constitutes the same amount of change in time whether it occurs in January, June, or December. Lastly, the most sophisticated level of measurement is the *ratio* level. In this level of measurement, the use of ratio (multiplicative) statements applied to the numerals assigned is meaningful for the underlying construct being measured.

Command Snippet 2.13: Creating and examining a subset of the `latino` data frame that only includes immigrants who came to the United States between the ages of 4 and 7.

```
## Create a subset of the data
> latino.sub <- subset(latino, subset = latino$ImmAge >= 4 &
    latino$ImmAge <= 7)

## Summaries of the object
> summary(latino.sub)
    Achieve            ImmYear              ImmAge
 Min.   :36.40    Min.    :20.60    Min.    :4.000
 1st Qu.:58.50    1st Qu.:64.00    1st Qu.:4.600
 Median :63.10    Median :72.00    Median :5.400
 Mean   :62.38    Mean    :67.74    Mean    :5.412
 3rd Qu.:68.90    3rd Qu.:76.70    3rd Qu.:6.100
 Max.   :85.80    Max.    :80.80    Max.    :7.000

 English            Mex
 Min.   :0.0000    Min.    :0.0000
 1st Qu.:1.0000    1st Qu.:1.0000
 Median :1.0000    Median :1.0000
 Mean   :0.8049    Mean    :0.7805
 3rd Qu.:1.0000    3rd Qu.:1.0000
 Max.   :1.0000    Max.    :1.0000
```

Variables that are at the ratio level of measurement always have an absolute zero. Weight, length, density, and resistance are all examples of ratio-level scales.

2.7.3 Categorical Data

Categorical, or nominal, variables have values that represent groups or categories. One of the variables in the `latino` data frame is such a variable. `English` categorizes immigrants by whether or not they are fluent in English. Those who are fluent in English are given the value 1, and those who are not are given the value 0. As with a number of categorical variables, `English` can be conceived of as an ordinal variable as its value indicates the extent of fluency. However, if the order of the categories is unimportant to the research questions—as it is here—then the variable is treated as categorical.

Categorical data is generally summarized using *contingency tables*, *bar plots*, or *pie charts*. Contingency tables are tabular representations of either the counts or proportions for each category, or label, represented in the categorical variable. The `table()` function can be used to count the number of cases in each category. The function requires at least one variable that can be interpreted as a categorical variable. Command Snippet 2.14 shows the syntax for creating a contingency table for the English fluency data from the `latino` data frame.

Command Snippet 2.14: Using the table() function to create a contingency table of the data in the English variable of the latino data frame. This is assigned to an object called my.tab which is then printed.

```
## Create contingency table
> my.tab <- table(latino$English)

## Print contingency table
> my.tab

  0   1
 38 112
```

In many cases, contingency tables are reported using proportions rather than the actual frequencies. The function prop.table() can be used on the assigned table to provide proportions in the contingency table. Command Snippet 2.15 shows the use of prop.table().

Command Snippet 2.15: Using the prop.table() function to create a contingency table of proportions for the data in the English variable of the latino data frame.

```
> prop.table(my.tab)

        0         1
0.2533333 0.7466667
```

In this case, almost 75% of the sample is fluent in English. The table() function can also take additional variables as arguments. When this is so, the contingency table will produce counts of the cross-classification of the variables included. For example, it may be of interest to know how many of the Latinos that immigrated from Mexico are fluent in English versus the number that immigrated from other countries that are fluent in English. Command Snippet 2.16 shows the syntax for creating a contingency table for the cross-classification of English fluency and whether or not the Latino immigrated from Mexico. The argument dnn= provides the dimension names to be used by supplying a vector of labels for the dimensions. Without this argument, the rows and columns are not labeled, and it must be remembered that the first variable supplied to table() will be the rows and the second variable will be the columns

Based on the contingency table, it can be seen that for the Latinos who immigrated from Mexico, 82 are fluent in English compared to 34 nonfluent English speakers. For Latinos who immigrated from other countries, the ratio of fluent to nonfluent speakers is even higher (30 : 4). One thing to note is that by summing the rows (which represent the English fluent Latinos), sums of 38 and 112, respectively, the same values are obtained as when the table() function is used on the English variable alone. Summing the columns gives the counts for the number of Latinos

Command Snippet 2.16: Using the `table()` function to create a contingency table of the data in the `English` variable of the `latino` data frame. This is assigned to an object called `my.tab.2` which is then called.

```
## Label the rows in the contingency table
## Create two-way contingency table and label dimensions
> my.tab.2 <- table(latino$English, latino$Mex, dnn =
    c("English", "Mex"))

## Print contingency table
> my.tab.2

      Mex
English  0  1
      0  4 34
      1 30 82

## Print contingency table of proportions
> prop.table(my.tab.2)

      Mex
English        0          1
      0 0.02666667 0.22666667
      1 0.20000000 0.54666667
```

who immigrated from Mexico and those that did not. These are sometimes referred to as the *marginal counts*.

It should be kept in mind that the `table()` function will *interpret* the variables provided in its arguments as categorical. Recall from the output of the `str()` function that both `English` and `Mex` are of type integer—even though they represent measurements at the nominal level. Data analysts need to be aware not only of the actual level of measurement for each variable in an analysis but also how R (or any software for that matter) is treating those variables. This helps an analyst make meaningful interpretations and inferences regarding the output that is produced.

2.7.4   Plotting Categorical Data

Categorical data are typically plotted using a bar plot, which is a graphical summarization of the same information produced in the contingency table.[12] The `barplot()` function in R will produce a bar for each of the interpreted categories from the `table()` function, where the height of each bar is proportional to the case count for each category. As such, a required argument for the `barplot()` function is the `height=` argument which takes the output from the `table()` function. Command

---

[12] Another common graph for displaying categorical data is the pie chart, which psychologists have pointed out is a very bad way of displaying information. It is an undesirable method as humans are not good at judging relative areas, especially areas produced in pie charts (e.g., Cleveland, 1993).

Snippet 2.17 shows the syntax for producing a bar plot for the English fluency data. The optional argument `names.arg=` can be used to provide a vector of names for the bars. The first command will produce a bar plot showing the frequencies of each category, and the second will produce a bar plot displaying the proportions. Figures 2.7 and 2.8 show the frequency bar plot and the proportion bar plot for the English fluency variable, respectively.

Command Snippet 2.17: Using the `barplot()` function to produce a bar plot of the data in the `English` variable of the `latino` data frame. The first command will produce a bar plot showing the frequencies of each category and the second will produce a bar plot displaying the proportions.

```
## Bar plot of the counts
> barplot(height = my.tab, names.arg = c("Not Fluent",
    "Fluent"))

## Bar plot of the proportions
> barplot(height = prop.table(my.tab), names.arg = c("Not
    Fluent", "Fluent"))
```

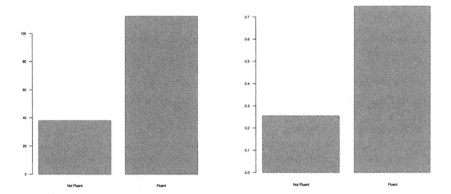

Figure 2.7: Bar plot showing the frequencies for Latinos in the sample who are fluent and not fluent in English.

Figure 2.8: Bar plot showing the proportions for Latinos in the sample who are fluent and not fluent in English.

2.8 FURTHER READING

The R documentation describes the import and export of data in many formats (R Development Core Team, 2009). Spector (2008) presents more detail on the importing of data into R from a variety of formats and also how to effectively manipulate that data. Zahn (2006) has written an article that describes the integration of the script and documentation file through a system called Sweave. This program can be used with R to write LATEX documents in American Psychological Association (APA) style, complete with results, tables, and figures.

Methodologically, the seminal paper by Stevens (1946) is a must-read for understanding the classification of levels of measurement. A more in-depth treatment can be found in Stevens (1951). Lord (1953) provides an amusing and satirical take on the same topic. A modern treatment is found in Long et al. (2003). Lastly, for more instruction and insight for analysis involving categorical data, the interested student is referred to Agresti (2002).

PROBLEMS

2.1   A data codebook is a printable file containing complete technical descriptive information for each variable in the corresponding data file. It describes how the data are arranged in the data file, what the various response values mean, a description of response-code labels, and any special instructions on how to use the data properly. Locate a data set (either one you have access to, a data set from a publication, or one found via the Internet) and create a codebook for this data set. The codebook should include:

- Description of the study including who carried out the study and what was their purpose.

- Sampling information: What was the population studied? How was the sample selected?

- Variable names.

- Variable values and labels.

- Any missing data codes.

2.2   Data were obtained from the New York State Department of Conservation (ozone data) and the National Weather Service (meteorological data) on daily air quality in New York City. The data are located in the R package **datasets** as a built-in data set called `airquality`. To read in a data set from this package we first need to load the **datasets** package using the `library()` function by typing `library(datasets)` at the prompt. The data set is now available in a data frame object called `airquality`. Typing the name of the object, `airquality`, at the prompt should print the data to the screen. A codebook for the data can be examined by typing `help(airquality)`.

   a) Produce a density plot of daily wind speeds. Describe the distribution by pointing out any interesting features of the plot.

   b) Using the `subset()` function, create an object that contains the air quality data for only the month of June. Use this new object to produce a density plot of solar radiation.

c) Produce a density plot for the solar radiation for the month of September. Compare and contrast the solar radiation for the month of June and the month of September.

d) This data set has missing values—as denoted by entries of NA in both the Ozone and Solar.R variables. How does the mean() function deal with missing values when finding the average?

e) At the prompt type mean(airquality$Ozone, na.rm = TRUE). How does the mean() function deal with NA values when the argument na.rm = TRUE is included?

CHAPTER 3

DATA EXPLORATION: ONE VARIABLE

Let the data speak for itself.

—J. W. Tukey (1977)

In 1986, the Vietnamese government began a policy of *doi moi* (renovation) and decided to move from a centrally planned command economy to a "market economy with socialist direction." As a result, Vietnam was able to evolve from near famine conditions in 1986 to a position as the world's third largest exporter of rice in the mid-1990s. Between 1992 and 1997 Vietnam's gross domestic product (GDP) rose by 8.9% annually (WorldBank, 1999).

The first Vietnam Living Standards Survey (VLSS) was conducted in 1992–93 by the State Planning Committee (SPC) (now Ministry of Planning and Investment) along with the General Statistical Office (GSO). The second VLSS was conducted by the GSO in 1997–98. The survey was part of the Living Standards Measurement Study (LSMS) that was conducted in a number of developing countries with technical assistance from the World Bank.

The second VLSS was designed to provide an up-to-date source of data on households to be used in policy design, monitoring of living standards, and evaluation of policies and programs. One part of the evaluation was whether the policies and

Comparing Groups: Randomization and Bootstrap Methods Using R
First Edition. By Andrew S. Zieffler, Jeffrey R. Harring, & Jeffrey D. Long

programs that were currently available were age appropriate for the population. For example, if a country has a higher proportion of older people, then there needs to be programs available that appeal to that sector of the population. Another concern was whether the living standards for different sections of the country were equitable.

Given the background above, data from the second VLSS (available in the VLSSage.csv and VLSSperCapita.csv data sets) is used to examine the following research questions:

1. What is the age distribution for the Vietnamese population?

2. Are there population differences in the annual household per capita expenditures between the rural and urban populations in Vietnam?

3. Are there population differences in the annual household per capita expenditures between the seven Vietnamese regions?

In the next few chapters we will use data exploration to answer these research questions. Data exploration is a key first step in any statistical analysis. It helps researchers discover both systematic structure in the data and also potentially problematic cases that deviate from that structure. Furthermore, good exploratory analysis can guide later analyses by suggesting appropriate models that may be investigated and also help the researcher to generate hypotheses not initially considered.

3.1 READING IN THE DATA

According to the codebook for the VLSSage.csv data, the variable Age contains the ages of 28,633 individuals (in years ranging from 0 to 99) living in the 5999 sampled households. To address the first research question regarding the age distribution, the data must be read in, and graphical and numerical summaries examined.

To read in tabular comma-delimited data with no missing values and variable names in the first row, the read.table() function is used with the arguments file=, header=TRUE, and sep=",". Remember that the file= argument takes a character string, which includes the entire path and filename of the data. The row.names= argument can also be included to assign the ID variable as the row names. Command Snippet 3.1 provides the syntax for reading in the VLSSage.csv data on a Mac (the file location is slightly different on a Windows machine).

After reading the data into R, a check should be made that it has been properly interpreted. It is good to examine the first few and last few rows of the data frame to be sure that it looks consistent with expectations. This is done by calling the head() and tail() functions on the assigned data frame object, respectively. The examination of the data frame object vlss is carried out in Command Snippet 3.1.

The output looks reasonable based on the codebook—there are two variables, and the first few cases look plausible for the variable on hand. If, for example, the ages had negative values, this would be evidence that something was amiss.

Secondly, the data structure—how R "sees" the data—should be inspected by calling the str() function on the assigned data. The str() output provides useful

Command Snippet 3.1: Read in and examine the VLSS age data. Note that the path provided in the file argument of the `read.table()` function is from a Mac.

```
## Read in the VLSS age data
> VLSS <- read.table(file = "/Documents/Data/VLSSage.csv",
    header = TRUE, sep = ",", row.names="ID")

## Examine the first six cases
> head(vlss)
   Age
1   68
2   70
3   31
4   28
5   22
6    7

## Examine the last six cases
> tail(vlss)
         Age
28628      6
28629     66
28630     48
28631     23
28632     19
28633     13

## Examine the data structure
> str(vlss)
'data.frame': 28633 obs. of  1 variable:
 $ Age: int   68 70 31 28 22 7 57 27 23 0 ...

## Summarize the data frame
> summary(vlss)
      Age
 Min.    : 0.00
 1st Qu.:12.00
 Median :23.00
 Mean    :28.14
 3rd Qu.:41.00
 Max.    :99.00
```

information, which is also seen in Command Snippet 3.1. The output indicates that there are 28,633 observations, and there are two variables. The last part of the output lists the first few values of each variable in the data frame and indicates the type or *class* of each variable. For example, R is storing each of these variables as integers (int). This is important to determine since certain functions can only be used on certain classes.

Lastly, basic statistics for each variable in the data frame are computed by using the summary() function on the data frame object. This also aids in the detection of entry errors and other strange values, etc. As seen in Command Snippet 3.1, the minimum and maximum values seem reasonable given the variable in question, age. These initial inspections suggest that we can move on to the analysis addressing the research question.

3.2   NONPARAMETRIC DENSITY ESTIMATION

Density estimation is one of the most useful methods for examining sample distributions. Nonparametric density estimation is an approach to estimating the distribution of the population from which the sample was drawn. This allows the observed sample data to provide insight into the structure of the unknown probability distribution underlying the population. The density estimation is nonparametric in the sense that the sample data suggest the shape, rather than imposing the shape of a know population distribution with particular values of its parameters.

3.2.1   Graphically Summarizing the Distribution

Graphically displaying the data allows a researcher to examine the overall patterns while also illuminating features or cases that are unexpected. For instance, there might be cases that are extreme relative to the majority of the data. For the VLSS data frame, we might have one or two very old people with ages greater than 100. It is important to be aware of such extreme cases as they might require special consideration in any statistical analysis to be performed.

3.2.2   Histograms

In the past, it was common for behavioral scientists to use a histogram to estimate the population density. A histogram is one of the simplest nonparametric estimators of a population density. Histograms, usually displayed as a graph, are essentially enumerations, or counts, of the observed data for each of a number of disjoint categories, or bins. Despite the popularity of the histogram, it has a number of drawbacks that suggest it is not always the best method to use.

The bin width—which is often chosen by the software rather than the researcher— has a great deal of influence on the shape of the histogram, and thus, on the inferences made by the researcher. Most methods for creating histograms partition the observed data into equally spaced bins using algorithms that typically focus on producing the optimal bin count given the observed data (Freedman & Diaconis, 1981; Scott, 1979; Sturges, 1926). These methods often have strong assumptions about the shape of the underlying population distribution.

For example, when distributions are strongly skewed, the density estimate produced by the histogram may be misleading, since several of the bins chosen in the optimization process can have very little data (e.g., Rissanen, Speed, & Yu, 1992).[1]

---

[1] Recently, Wand (1997) proposed a series of "plug-in" rules for selecting the bin width in a histogram. The resulting histogram provides a better estimate for the density than other methods.

Perhaps most importantly, histograms are essentially discontinuous step functions. A discontinuous function is one in which there are clearly gaps when the graph of the function jumps to each new x-value. So, if the researcher believes that the observed sample data comes from a population with a continuous density, then other estimation methods are preferable. Fortunately, there are substantially better methods of estimating the population density.

3.2.3 Kernel Density Estimators

A much better estimate of the population density can be obtained by using kernel methods. Nonparametric kernel density estimation can be thought of as a method of averaging and smoothing the density estimate provided by the histogram. More formally, kernel density estimation is a sophisticated form of locally weighted averaging of the sample distribution.

Figure 3.1 shows a conceptual illustration of kernel density estimation (adapted from Sain (1994)). The vertical lines below the axis represent the $N = 6$ sample observations. The dashed lines represent the Gaussian kernel function, and the solid line represents the overall density estimate. The smoothing parameter, represented by the variation in each distribution, is fixed (i.e., it is the same value in each kernel).

Kernel density estimation works by estimating the density at each observation, x, using a smooth, weighted function, known as a *kernel*. In Figure 3.1 the kernel function is the normal, or Gaussian, distribution, however, there are several kernel functions that can be used in density estimation. Each kernel function is centered at one of the $N = 6$ observations and identically scaled. An estimate of the overall density can then be found by summing the height of the kernel densities at each observation.

Note that the variation (width) in the kernel determines the amount of overlap at each observed value. Skinnier kernels have less overlap resulting in a smaller overall sum, while wider kernels would result in more overlap and a larger sum. The data analyst not only specifies the kernel function, but also the variation in the kernel function. Figure 3.1 shows a visual depiction of the variation in the kernel functions based on the half-width (i.e., half the width of each kernel) of the kernel. This is also referred to as the *smoothing parameter* since it has a direct impact on the overall smoothness of the plot.

3.2.4 Controlling the Density Estimation

Based on Figure 3.1, changing either the kernel function or the smoothing parameter would affect the overall density that is estimated. There are several kernel functions that can be used in R. While there are differences in the shape of the kernel functions, these differences are negligible. The research literature suggests that the choice of the functional form of the kernel has little effect on the estimation of the density—all are essentially equivalently efficient in minimizing the error when approximating the true density (e.g., Epanechnikov, 1969; Silverman, 1986)—and thus the default Gaussian kernel function is often used.

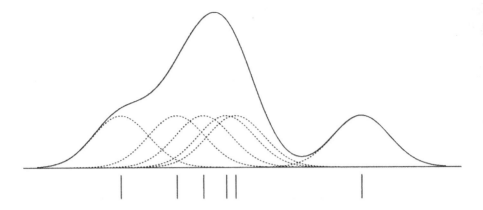

Figure 3.1: Illustration of the kernel density estimation (solid line) for $N = 6$ observations (vertical lines). A Gaussian kernel function (dashed lines) with a fixed smoothing parameter was centered at each observation. The figure was adapted from Sain (1994).

Selecting the smoothing parameter is another matter. Change in the amount of variation in the kernel has a large effect on the appearance and interpretation of the density estimate. This is similar to forcing fewer or additional bins in the traditional histogram. Figure 3.2 shows the difference in appearance that is associated with changing the smoothing parameter. The three density estimates are based on the same sample of data, and all use the Gaussian kernel function. It can be seen that using a smoothing parameter of 0.5 produces an estimate of the density which is quite rough (left-hand plot), while using a smoothing parameter of 2 (middle plot) produces a smoother estimate. A smoothing parameter of 10 (right-hand plot) produces an even smoother plot. As the smoothing parameter increases, the density curve becomes smoother.

If the researcher chooses a smoothing parameter that is too small, the density estimate will appear jagged, spuriously highlighting anomalies of the data such as asymmetry and multiple modes. Such features can appear because of chance variation rather than because they are structures present in the probability distribution. If the researcher chooses a smoothing parameter that is too large, she may obscure much of the structure in the data, a phenomenon known as oversmoothing. Ideally a smoothing parameter is chosen that is small enough to reveal detail in the graph but large enough to inhibit random noise.

Several methods have been proposed to choose an optimum smoothing parameter based on the data (see Sheather, 2004). While these methods tend to compute smoothing parameters that perform well in simulation studies, for sample data that is substantially nonnormal, some manual adjustment may be required. Another method is the use of *adaptive kernel estimation* (e.g., Bowman & Azzalini, 1997; Silverman, 1986; Terrell & Scott, 1992), especially for estimating densities of long-

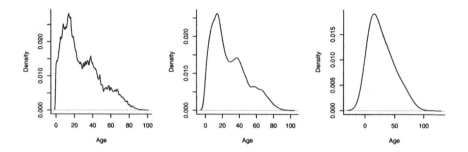

Figure 3.2: Kernel density estimates using three different smoothing parameters. Each estimate used a Gaussian kernel. The estimate on the left used a smoothing parameter of 0.5. The estimate in the middle used a smoothing parameter of 2. The estimate on the right used a smoothing parameter of 10.

tailed or multi-modal data. Adaptive kernel density estimation allows the kernels to have differing smoothing parameters. Varying the smoothing parameter reduces the potential of undersmoothing the density estimate where there are sparse data and also the potential of oversmoothing the density estimate where there are heavy data.

Unfortunately, there is no one uniform best method for choosing the smoothing parameter (Simonoff, 1996). Terrell (1990, p. 470) asserts that "most density estimates are presumably designed on aesthetic grounds: The practitioner picks the smoothing parameter so that the density looks good. Clearly, though, such an individualistic approach does not lead to replicable results; nor does it lead reliably to sensible estimates from a novice."

3.2.5 Plotting the Estimated Density

To obtain a kernel density estimate, the density() function is used. The function is provided a vector of data via the argument x=. For the example, kernel density estimation is performed for the Age variable from the VLSS data frame. In order for R to recognize Age, it must be associated with the VLSS data frame. This is accomplished by using the $ operator. The $ operator allows the extraction of one variable, or access to one column of a data frame. Think of the data frame name as the "family name" (e.g., Smith) and the variable name as the "given name" (e.g., John). Putting these together yields Smith$John.

The density() function can also be supplied with the optional argument bw= to adjust the smoothing parameter.[2] The output can be assigned to an object, say d, which will store the estimated kernel density estimates. The density() function produces density estimates for 512 equally spaced points by default. This object can then be printed to provide summary information about the density estimates. For a complete list of these values, the command d$x or d$y can be issued.

Command Snippet 3.2: Syntax to estimate the density of the Age variable. The density object is then printed and plotted.

```
## Estimate the density
> d <- density(vlss$Age)

## Print the density object d
> d
Call:
  density.default(vlss$Age)

Data: VLSS$Age (28633 obs.);  Bandwidth 'bw' = 2.344

        x                    y
 Min.   : -7.033    Min.    :1.008e-07
 1st Qu.: 21.234    1st Qu.:1.410e-03
 Median : 49.500    Median :6.416e-03
 Mean   : 49.500    Mean    :8.836e-03
 3rd Qu.: 77.766    3rd Qu.:1.385e-02
 Max.   :106.033    Max.    :2.587e-02

## Plot the density object d
> plot(d)
```

Command Snippet 3.2 shows the syntax used to assign, print, and plot the estimated density of Age. To plot the kernel density estimate, the plot() function is used on the estimated densities, which are stored in the density object d. The plot() function draws a line graph of the density based on the x and y values computed by the density() function. The left-hand plot of Figure 3.3 shows the result of calling the plot() function on the density object d.

In the example, the computed value of the smoothing parameter is 2.344 (see Command Snippet 3.2).[3] This value can be obtained by printing the density object (see Command Snippet 3.2). Although the default smoothing parameter in this case seems reasonable in providing a smooth plot, the prudent data analyst should try several smoothing parameters to ensure that the plot is not oversmoothed. Figure 3.3 shows the original plot, as well as additional kernel density plots using the Sheather

---

[2]The argument kernel= can also be specified.

[3]The default value for the smoothing parameter is selected using Silverman's rule of thumb.The method popularized by Silverman—which was originally proposed for improved density estimation in histograms (e.g., Deheuvels, 1977; Scott, 1979)—is a computationally simple method for choosing the smoothing parameter. See Silverman (1986) for more details.

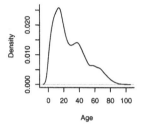 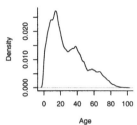 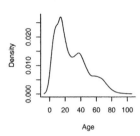

Figure 3.3: Kernel density estimates using Silverman's rule of thumb for the smoothing parameter (left-hand plot). This is the default value for the smoothing parameter. The middle plot uses the Sheather and Jones "solve the equation" method of computing the smoothing parameter, and the right-hand plot uses an adaptive kernel function.

and Jones "solve the equation" method of computing the smoothing parameter and an adaptive kernel function. All the methods yield similar curves for the Age variable.

The second plot in Figure 3.3 can be created by adding the argument bw="SJ" to the density() function (see Command Snippet 3.3). The argument bw= takes either a numerical value to specify the smoothing parameter or a character string that identifies the method that should be used to compute the smoothing parameter. For example, bw="SJ" will compute the smoothing parameter based on the Sheather and Jones "solve the equation" method. The plots in Figure 3.2 were created by using bw=0.5, bw=2, and bw=10.

Command Snippet 3.3 also includes the syntax to estimate the density of the ages using an adaptive kernel function. This is estimated using the akj() function which is in the **quantreg** package. The vector that provides the data used for estimating the density is provided to the x= argument. The z= argument takes an equispace sequence of values for which the density should be estimated. From the output in Command Snippet 3.2, recall that the density object, d, contains a set of equispaced x-values at which the density was estimated. These can be accessed using d$x, which is then supplied to the z= argument.

The estimated density is assigned to the object d.akd. Because of the iterative method that adaptive kernel estimation uses, running the function can take some time depending on the computer's processor and the size of the sample. Lastly, the density is plotted using the plot() function. Because the values for x and y, that are to be plotted, are contained within different objects, we must specify this in the plot() function. The argument type="l" draws a line plot rather than just plotting the ordered pairs of x and y.

In all of the plots, the estimated kernel density shows that the population has three potential modes in the distribution. These likely refer to three distinct age groups or subpopulations. Each of these age groups has a higher density (taller distribution)

Command Snippet 3.3: Changing the smoothing parameter in the density plot.

```
## Density plot using smoothing parameter from Sheather and
     Jones method
> d.SJ <- density(vlss$Age, bw = "SJ")
> plot(d.SJ)

## Density plot using adaptive kernel density estimation
> library(quantreg)
> d.akd <- akj(x = vlss$Age, z = d$x)
> plot(x = age, y = d.akd$dens, type = "l")
```

and seems to have less variation (a thinner distribution) than the subsequent younger age group.

3.3   SUMMARIZING THE FINDINGS

Researchers have many options when presenting their findings. Statistical analyses are presented in tables, graphs, and/or prose. Statistics texts sometimes provide little guidance for such reporting. This is likely due to the fact that reporting styles vary widely based on the outlet for the work. What is appropriate for reporting in a master's thesis might be inappropriate for a professional journal and vice versa. Throughout this monograph we will try to offer some guidance using the standards put forth by the American Psychological Association (APA; American Psychological Association, 2009).

When summarizing the findings, it is desirable to synthesize the statistical information culled from the data analysis with the content and findings from the substantive literature. This helps the researcher either verify or call into question what she is seeing from the analysis. Relating the findings back to previous research helps the researcher evaluate her findings. The results might verify other findings or question them. The researcher might also question her own results if they are not consistent with expectations and other studies. There are two things to keep in mind:

- There is no "right" answer in statistical analysis (Box & Draper, 1987; Box, 1979) but only explanation and presentation.

- There is a distinction between how most researchers do their analysis and how they present their results.

With the accumulation of statistics knowledge comes the prospect of a greater number of choices in the data analysis. Different researchers may make different analytic decisions; and while, hopefully, the substantive findings about the research questions will not change, it is possible that they may. Furthermore, different researchers will choose to make different decisions about what information to present in a paper. When presenting findings, regardless of the decisions made or analytic

approach taken, there needs to be sufficient information included to judge the sound-
ness of the conclusions drawn. Section 3.2.2 provides an example of what constitutes
sufficient information in the write-up of the results.

3.3.1   Creating a Plot for Publication

The *Publication Manual of the American Psychological Association* (American Psychological Association, 2009) is judicious in the encouragement of the use of graphs
and figures. However, figures are only appropriate when they complement the text
and can eliminate a lengthy textual discussion. According to the APA, a good graph
should have, "lines [that] are smooth and sharp; typeface [that] is ... legible; units
of measure are provided; axes are clearly labeled; and elements within the figure are
labeled or explained" (p. 153). R has functionality that allows the user to produce
highly customized graphs. One can add labels, titles, legends, etc., very easily to
almost any plot. R can be used to make a plot of publishable quality very quickly.

The main function for creating graphs discussed in this monograph is the plot()
function. The plot() function includes several optional arguments that can be
included to create a figure that meets the APA standards. Below, some of these
arguments are illustrated to create a publication-quality figure displaying the density
of the age distribution for the VLSS data. The syntax used is provided in Command
Snippet 3.4, whereas the plot itself is shown in Figure 3.4.

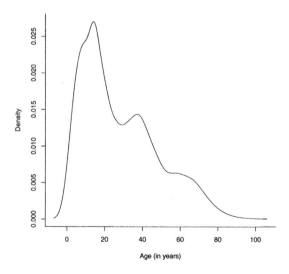

Figure 3.4: Kernel density estimate for the distribution of ages for a sample of 28,633
Vietnamese citizens using an adaptive kernel method.

Command Snippet 3.4: Code to plot the age data for publication.

```
## Add axes labels
> plot(x = d$x, z = dakd$dens, type = "l", xlab = "Age (in
      years)", ylab = "Density")

## Remove main title
> plot(x = d$x, z = dakd$dens, type = "l", xlab = "Age (in
      years)", ylab = "Density", main = " ")

## Change box type
> plot(x = d$x, z = dakd$dens, type = "l", xlab = "Age (in
      years)", ylab = "Density", main = " ", bty = "l")
```

Labels   Axes labels should be informative to the reader and include units of measure if they aren't readily apparent to the reader. The default *x*-axis label for the density plot does not represent the *x*-axis scale, a problem that should be corrected. For the figure, it is desirable to have a *x*-axis label of *Age (in years)*. The default *y*-axis label is fine. According to APA, axes labels should use text in "heading caps" parallel to the axis. For example, the *y*-axis label should be readable when the graph is rotated 90 degrees clockwise. The arguments xlab= and ylab= can be added to the plot() function to change the axes labels. Both arguments require a quoted character string, which consists of the desired label.

Captions   Although the title of the plot can be changed using the main= argument, when preparing a figure for an APA journal, a plot title should not be included (omit main=). Rather, a figure caption should be created in the word-processing document. Sometimes R puts in a title by default, in which case it may be necessary to remove it. By providing an empty character string, main=" ", it is possible to remove any default title that R might input.

APA has specific guidelines pertaining to figure captions. Figure captions are placed below the figure and use Arabic numerals in italics to define them (e.g., *Figure 1.*). Since the caption "serves both as an explanation of the figure and as a figure title" (American Psychological Association, 2009, p. 159), it needs to be descriptive yet concise. The caption should also include any information a reader needs to interpret the figure ... "a reader should not have to refer to the text to decipher the figure's message" (p. 160). Captions are printed using standard typeface and end with a period. For example, the title and caption for our plot might read

Figure 1. Kernel density estimate for the distribution of ages for a sample of 28,633 Vietnamese citizens using an adaptive kernel method.

Axes   Some professional journals require there be no box around the plotting area. That is, only the *x*-axis and *y*-axis are drawn. This practice is encouraged by the APA. The box type around the plot can be controlled by including the bty= argument to the `plot()` function. The argument `bty="l"` is used to remove the top and right side of the box so that only the *x*- and *y*-axes are included in the plot.[4]

The process suggested by the syntax in Command Snippet 3.4 is to *build* a plot, piece by piece, saving code that produces the desired effects in the script file. For example, one could initially start with a density plot of the ages, then add labels and a title. There are many additional changes that can be made. Each time additional syntax is executed, the plot feature is added to an existing graph. An error message will be produced when there is an attempt to add a feature when there is no existing graph. To remake an entire graph, the initial `plot()` command must be re-run. If a command produces an undesirable result, it is easy to re-create the graph from the beginning using alternative features by revising the syntax in the script file.

3.3.2 Writing Up the Results for Publication

When presenting results, researchers should integrate or synthesize information from the data analysis with the content and findings from the substantive literature. The writing should also be clear and concise. Since many journals have a page or word limit, precious space should not be used up presenting every nuance of the statistical results. Rather, focus should be on the results most relevant to the research questions. Remember, the statistical analysis is used to inform the substantive area, and the writing should reflect this. An example of a write-up for the age distribution analysis might be as follows.

3.4 EXTENSION: VARIABILITY BANDS FOR KERNEL DENSITIES

One concern that is raised by using the density estimate in exploring univariate data is the need to evaluate which features of the density estimate indicate legitimate underlying structure and which can be attributable to random variation. For example, does the Age variable in the VLSS data frame actually contain three modes? Or, in particular, is the third mode attributable to chance variation produced by poor estimation of the density at those age values? One method of capturing the uncertainty associated with a density estimate is to create a *variability band* or *interval* around the density estimate. As demonstrated by Bowman and Azzalini (1997), variability bands around the density estimate, say, $\hat{p}(x)$, have the form

$$\hat{p}(x) \pm 2 \times \text{SE}_{\hat{p}(x)}, \tag{3.1}$$

where $\text{SE}_{\hat{p}(x)}$ is the standard error computed from the variance estimate, which in turn is computed using a Taylor series approximation (see Bowman & Azzalini, 1997, for further details).

---

[4]Different boxes are drawn by using various character strings in the `bty=` argument. We use `"l"` since the box we want—the left and bottom sides of a box—resemble and correspond to the letter L depicted on the keyboard. For other options, use `help(par)`.

Sample Write-Up

The density for a sample of ages from $N = 28,633$ Vietnamese citizens was estimated using an adaptive kernel method (Portnoy & Koenker, 1989). The plot of the estimated kernel density reveals three potential subpopulations in the distribution which depict three distinct age groups: a younger, middle-aged, and older generation. The population pyramid based on the 1999 Vietnam census also seems to indicate the presence of three subpopulations (General Statistical Office, 2001). The overall positive skew in the distribution also suggests that the population of each subsequently higher age group has relatively lower frequency. Two other interesting features emerge in the density plot that have also been documented in the research literature (e.g., Haughton, Haughton, & Phong, 2001). The second mode in the age distribution occurs near the age of 40. This mode has lower frequency in part because of lives lost during the Vietnam War, which ended in 1975 with the reunification of North and South Vietnam. One can also see the effect of decreasing fertility rates in the fact that estimated density decreases as ages decrease from about 16 to zero.

Bowman and Azzalini (1997) have implemented functionality to compute and plot the density estimate along with variability bands in R, using the `sm.density()` function from the **sm** package. For illustrative purposes, consider a random sample of $N = 1000$, selected without replacement using the `sample()` function. Figure 3.5 shows the plotted density and variability bands for the subset of ages, using a smoothing parameter $h = 4$.

Bowman and Azzalini (1997) show that the density estimate of the true underlying density is biased[5] and the bias is a function of both the curvature of the estimate and the value of the smoothing parameter, h. In particular, $\hat{p}(x)$ underestimates $p(x)$ at the peaks in the true density and overestimates it at the troughs. As h increases, bias increases but variability decreases. As h decreases the opposite relation occurs.

Wand and Jones (1995) argue that the graphical display of the variance structure can be helpful when determining if interesting features of the density are indeed part of the underlying structure in the population. For example, the width of bands for the *VLSS* data strengthens the evidence that the troughs and peaks of the small wiggles seen in Figure 3.5 stay within each other's variability bands suggesting that there is no good evidence that the wiggles exist in the population. On the other hand, the different plateau near ages 35 and 70 are clearly distinct.

Command Snippet 3.5 shows the syntax used to create Figure 3.5. The argument `display="se"` in the `sm.density()` function indicates that variability bands are to

---

[5]Bias is defined as the difference being the true population density $p(x)$ and its nonparametric estimate $\hat{p}(x)$.

be plotted along with the density estimate. The argument h= indicates the smoothing parameter to be used in the density estimation.

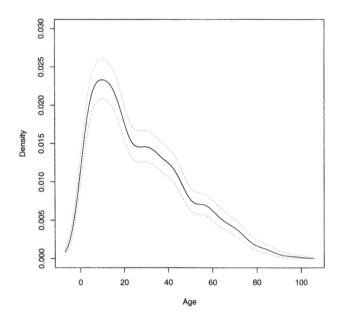

Figure 3.5: Kernel density estimate (solid line) with variability bands (dotted lines) for the distribution of ages for the Vietnamese population. A smoothing parameter of $h = 4$ was used in the density estimation.

Command Snippet 3.5: Code to plot the kernel density estimate of the age data along with variability bands.

```
> age.sample <- sample(vlss$Age, n = 1000, replace = FALSE)
> library(sm)
> sm.density(age.sample, h = 4, xlab = "Age", ylab = "Density",
     xlim = c(-7,106), ylim = c(0, 0.03), display = "se",
     rugplot = FALSE)
```

3.5 FURTHER READING

This section provides some references for the reader interested in building on the material presented in this chapter. While this list is not complete, it does offer some starting points. The philosophies underlying exploratory data analysis were introduced by Tukey (1962, 1977). Many other statisticians have written papers building on and implementing Tukey's initial ideas including Cleveland (1984), Dasu and Johnson (2003), and Leinhardt and Leinhardt (1980). Regarding the ideas and methods related to density estimation, the interested reader is referred to Bowman and Azzalini (1997), Sheather (2004), and Wilcox (2004). Overviews and surveys of the methods used to select a smoothing parameter are offered in Turlach (1991) and Jones, Marron, and Sheather (1996). For further guidance on presenting results, resources include American Educational Research Association (2006), American Psychological Association (2010), and Cummins (2009).

PROBLEMS

3.1 Using the VLSSperCaptia.csv data set, create a plot of the kernel density estimate for the marginal distribution of household per capita expenditures. Explore the effect of changing the smoothing parameter to determine whether there is evidence that this distribution is multimodal. Write up the results of your investigation as if you were writing a manuscript for publication in a flagship journal in your substantive area. Be sure to include a density plot of per capita expenditures using the smoothing parameter that you have settled on.

3.2 Use the following commands to draw 50 randomly selected values from a standard normal distribution—a normal distribution having a mean of 0 and a standard deviation of 1. (The line set.seed(100) will set the starting seed for the random number generation so that the same set of random observations will be drawn.)

```
> set.seed(100)
> x <- rnorm(n = 50, mean = 0, sd = 1)
```

    a) Create a density plot of x using a smoothing parameter of 0.5.

    b) The function density() has an optional argument, kernel= that allows density estimates to be constructed from different shapes of kernel functions. The default kernel function, when this argument is not included, is to use a Gaussian density estimate (kernel = "gaussian"). Using the same smoothing parameter of 0.5, create a density plot of x using each of the following kernel shapes: kernel = "triangular", kernel = "rectangular", kernel = "epanechnikov", and kernel = "biweight". (Note: You should have four different plots.)

    c) Describe the similarities *and* differences between the four plots.

3.3 Generate 100 values from a standard normal distribution and assign them to an object called y.

    a) Use the function sm.density() to create a density plot of y.

    b) Create a density plot of y, but this time include the argument model = "Normal" in the sm.density() function.

c) When the argument `model = "Normal"` is included in the function, a reference band, indicating where a density estimate is likely to lie when the data are normally distributed, will be superimposed on the plot. Based on this plot, is there evidence to suggest that the generated data are non-normal? Explain.

d) Repeat the same exercise, but this time draw a random sample of 100 observations from a chi-square distribution (skewed to the right) with 3 degrees of freedom. The syntax to draw these data is `z <- rchisq(n=100, df=3)`. Create a density plot of `z`. Based on this plot, is there evidence to suggest that the generated data are nonnormal? Explain.

CHAPTER 4

EXPLORATION OF MULTIVARIATE DATA: COMPARING TWO GROUPS

The nature of doing science, be it natural or social, inevitably calls for comparison. Statistical methods are at the heart of such comparison, for they not only help us gain understanding of the world around us but often define how our research is to be carried out.

—T. F. Liao (2002)

In Chapter 3 the Vietnam Living Standards Survey (VLSS) was introduced. The survey was designed to provide an up-to-date source of data on households to be used in public policy formation, to assess current living standards, and to evaluate the impact of public programs. The data set is used to address three research questions.

1. What is the age distribution for the Vietnamese population?

2. Are there differences in the annual household per capita expenditures between the rural and urban populations in Vietnam?

3. Are there differences in the annual household per capita expenditures between the seven Vietnamese regions?

Comparing Groups: Randomization and Bootstrap Methods Using R   **67**
First Edition. By Andrew S. Zieffler, Jeffrey R. Harring, & Jeffrey D. Long
Copyright © 2011 John Wiley & Sons, Inc.

The first research question was addressed in Chapter 3. In this chapter, focus is on whether the living standards for urban and rural Vietnamese are equitable and related questions about differences between the two groups.

4.1   GRAPHICALLY SUMMARIZING THE MARGINAL DISTRIBUTION

According to the codebook, the data set *VLSSperCapita.csv* contains data on the household per capita expenditures for $N = 5999$ households along with two demographic variables. Assuming that a new R session is started, we will again need to read in the data. The structure of the data is like the structure of the data in the *VLSSage.csv* file considered previously, namely tabular. Recall that tabular data is a cases (rows) by variables (columns) array. The variable names are included in the first row, each of the values is separated by a comma, and there are no missing data.

The data are read into R using the `read.table()` function. Command Snippet 4.1 provides an example. In the command snippet, after the data are read in, the data frame is examined using `head()`, `tail()`, `str()`, and `summary()`. The output looks reasonable based on the codebook—there are four variables, and the first few cases look plausible for all four variables.

Initially, the density estimate of the per capita expenditures for all 5999 households—ignoring whether they are urban or rural—is examined. The distribution ignoring the population area is known as the *marginal distribution*. Recall that in the last chapter, marginal density plots were created using the `density()` and `plot()` functions. The syntax to plot the density estimate for the marginal distribution of the per capita expenditures is shown in Command Snippet 4.2.

The plot of the marginal distribution shown in Figure 4.1 indicates that the per capita expenditures data is right skewed, as the majority of dollar amounts pile up at the low end, and taper off moving to the right. This suggests that many of the households in Vietnam have a low per capita expenditure (around $100 U.S.). It also shows some households that may be potential outliers in the marginal distribution. These are households with very high expenditures relative to the rest of the households in the sample.

4.2   GRAPHICALLY SUMMARIZING CONDITIONAL DISTRIBUTIONS

Examining the marginal distribution is useful in an initial examination of the data, but it does not help in answering the research question about rural and urban differences. To help address the research question, the distribution of per capita expenditures for each area must be examined separately. The distributions of per capita household expenditures for each area are called *conditional distributions* because they are defined conditional on area.

To graphically examine the conditional distributions, two new variables are created, one that contains the household per capita expenditures for the rural households and one for the urban households. The per capita expenditures for these two variables are plotted separately. Since there is a factor variable that specifies whether each

Command Snippet 4.1: Syntax to read in the per capita expenditure data.

```
## Read in the data
> household <- read.table(file =
    "/Documents/Data/VLSSperCapita.csv", header = TRUE, sep =
    ",", row.names = "ID")

## Examine the first part of the data frame
> head(household)
  Dollars  Area Region
1  184.33 Rural      5
2   62.73 Rural      5
3  119.13 Rural      5
4   76.61 Rural      5
5   97.46 Rural      5
6  132.09 Rural      5


## Examine the last part of the data frame
> tail(household)
      Dollars  Area Region
5994   320.83 Urban      3
5995   351.78 Urban      3
5996   119.21 Urban      3
5997   298.80 Urban      3
5998   303.77 Urban      3
5999   652.42 Urban      3

## Examine the structure of the data frame
> str(household)
'data.frame': 5999 obs. of  3 variables:
 $ Dollars: num  184.3 62.7 119.1 76.6 97.5 ...
 $ Area   : Factor w/ 2 levels "Rural","Urban": 1 1 1 1 1 1 1 1
     1 1 ...
 $ Region : int  5 5 5 5 5 5 5 5 5 5 ...

## Summarize the data frame
> summary(household)
    Dollars            Area          Region
 Min.   :   23.82   Rural:4269   Min.   :1.000
 1st Qu.:  111.41   Urban:1730   1st Qu.:3.000
 Median :  159.80                Median :5.000
 Mean   :  212.58                Mean   :4.361
 3rd Qu.:  247.40                3rd Qu.:6.000
 Max.   : 3053.45                Max.   :7.000
```

household in the sample is urban or rural (i.e., Area), the *indexing* operations in R are used to subset the data. The indexing operations in R are quite powerful and allow access to subsets of cases, or subsets of variables, or both, from the original data frame.

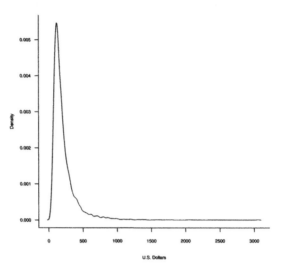

Figure 4.1: Kernel density estimate for the marginal distribution of household per capita expenditures (in U.S. dollars).

Command Snippet 4.2: Syntax to plot the marginal per capita household expenditures.

```
> plot(density(household$Dollars), xlab = "U.S. Dollars", main
    = " ")
```

4.2.1 Indexing: Accessing Individuals or Subsets

An element of a matrix or data frame is accessed using the bracket notation, x[i, j], where x is the name of the matrix or data frame, i is the row number, and j is the column number. The household data frame has 5999 rows and 5 columns, or 5999 × 5. To access the first household's per capita expenditure, the following syntax is used, household[1, 2]. Indexing is not limited to single elements; entire rows or columns can be selected or subsetted. To select an entire row or column, the appropriate index location is left blank. For example, x[1 ,] selects the first row and all the columns. The syntax x[,] selects all the rows and all the columns. The syntax x[,] can be shortened to x, which is convenient in many applications. Command Snippet 4.3 provides examples of syntax for indexing elements and vectors from the household data frame and elements from the vector household$Dollars.

Command Snippet 4.3: Indexing elements from the household data frame.

```
## Access the first household's per capita expenditure.
> household[1, 1]
[1] 184.33

## Access all the variables for the first household.
> household[1, ]
      ID Dollars Urban          Region Urban.Factor
1 12225  184.33     0 Northern Uplands            0

## Access every household's per capita expenditure.
## Note that the output is suppressed.
> household[ ,1]

## Access the first household's per capita expenditure.
> household$Dollars[1]
[1] 184.33
```

4.2.2 Indexing Using a Logical Expression

The first objective is to obtain the household per capita expenditures for only the rural households. Rather than select a single element from the household$Dollars vector, we will select those households for which Area is Rural. Command Snippet 4.4 shows the syntax for obtaining the household per capita expenditures for the rural and urban households, respectively, and assigning those values to new objects.

Command Snippet 4.4: Indexing using a logical expression.

```
> rural.households <- household$Dollars[household$Area ==
  "Rural"]
> urban.households <- household$Dollars[household$Area ==
  "Urban"]
```

Notice the use of *two* equal signs in the indexing in Command Snippet 4.4. It would seem that one or two equal signs should not make for a big difference, but to a program like R, it signifies a great deal of difference. A double equal sign indicates a *Boolean* or *logical* expression in R. A logical expression is an expression that is either evaluated as TRUE or FALSE.

The logical expression in the indexing is essentially asking if the value of household$Area "is equal to" Rural. If so, the logical expression is evaluated as TRUE and the value of household$Dollars for that household should be included in the new variable. If the logical expression is evaluated as FALSE, the household's per capita expenditure should not be included. Quotation marks must appear around Rural since it is a character string. Likewise, a variable that only contains urban households can be created using the logical expression Area=="Urban".

Command Snippet 4.4 shows the syntax for separately creating variables containing the household per capita expenditures for both the rural and urban households.

4.2.3   Density Plots of the Conditional Distributions

A graph of the density estimate for the rural households can be constructed by using the density() function with rural.households. Then the plot() function is used to plot the density of the per capita expenditures for the rural households. The same thing is done for the urban households. For purposes of comparisons, it is useful to draw both densities on the same plot. The plot() function does not allow for this. Rather, the lines() function must be used to add additional densities to an existing plot.

A density plot of the household per capita expenditures for the urban households is added to the existing plot of the rural households. To differentiate between the two densities, different types of lines are used, a solid line for the rural density and a dotted line for the urban density. The argument lty="solid" is used with plot() for the rural density curve and lty="dotted" is used with lines() for the urban density curve. The effect of the lines() function is to add the new density to the existing plot. Command Snippet 4.5 shows the syntax for plotting the densities for both conditional distributions on the same graph.

Command Snippet 4.5: Plotting the density curves for the rural and urban households.

```
## Plot density curve for the rural households
> d.rural <- density(rural.households)
> plot(d.rural, main = " ", xlab = "Household Per Capita
    Expenditures (in U. S. Dollars)", lty = "solid", bty = "l")

## Add density curve for the urban households
> d.urban <- density(urban.households)
> lines(d.urban, lty = "dotted")
```

Figure 4.2 shows a single graph—or panel—with the conditional density curves superimposed and coded by line type. By having both conditional distributions in the same panel, this type of plot makes it psychologically easier for people to make comparisons. As can be seen in the plot, the urban curve is shifted to the right of the rural curve toward higher dollar amounts. In addition, the peak of the urban curve is lower than that of the rural curve, and both distributions are positively skewed.

4.2.4   Side-by-Side Box-and-Whiskers Plots

Another graph that is useful for making comparisons is the side-by-side box-and-whiskers plot. The function to create a side-by-side box-and-whiskers plot is boxplot(). To produce side-by-side box-and-whiskers plots, the boxplot() function is supplied with the name of each variable to be plotted, separated by commas.

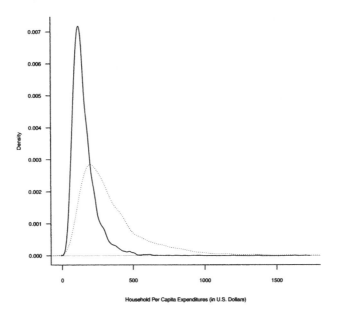

Figure 4.2: Kernel density plots for the distribution of household per capita expenditures (in dollars) conditioned on area. The density of household per capita expenditures for urban households is plotted using a solid line and that for rural households is plotted using a dotted line.

The optional argument names= is used to produce better labels for each box plot. Command Snippet 4.6 shows the syntax used to create the side-by-side graphs in Figure 4.3. If the text labels are long, the plot may have to be resized to accommodate them.

Command Snippet 4.6: Syntax to plot the distribution of per capita household expenditures conditioned on area using a side-by-side box-and-whiskers plot.

```
> boxplot(rural.households, urban.households, names =
    c("Rural", "Urban"))
```

Box-and-whiskers plots provide a flexible and effective manner in which to graphically summarize the distributions of the variables. They highlight the center (median) of the data, the variation in the central part of the data, how the tails relate to the central part of the data, and they display *potential outliers*. The box and the "fences" connected to it provide an indication of the asymmetry and variation in the middle part of the distribution. The horizontal line in the box represents the median. The length of the box provides an indication of the range for the central 50% of the scores in the distribution. When the median is close to the lower box edge, this indicates a positively skewed distribution. The box-and-whiskers plot for rural in Figure 4.3 is an example. When the median is close to the upper box edge, this indicates a negatively skewed distribution. When the median line is roughly in the center of the box, then the middle 50% of the distribution is relatively symmetric. The box-and-whiskers plot for urban in Figure 4.3 is an example, as the median is approximately equidistant from the box edges. However, the numerous points in the upper portion indicate the overall distribution is not symmetry, only the middle 50%.

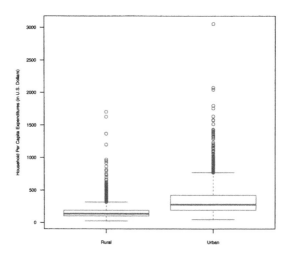

Figure 4.3: Side-by-side box-and-whiskers plots for the conditional distributions of household per capita expenditures (in U.S. dollars) using the `names=` argument.

4.3   NUMERICAL SUMMARIES OF DATA: ESTIMATES OF THE POPULATION PARAMETERS

After graphically examining the data, it is desirable to obtain a more precise numerical summarization of the estimated population distribution. The numerical summaries can generally be split into two different types:

- Measures of location, or central tendency
- Measures of variability, or dispersion

Measures of location are single values that represent the measurement of a typical individual or unit in the distribution being studied. For example, in Figure 4.3, a typical household in the distribution might be defined as having a per capita expenditure at the dollar amount directly below the peak of the curve. Based on this, the typical urban household has a higher per capita expenditure than the typical rural household.

Measures of variability provide an indication of how different, or variable, the measurements in the distribution happen to be. For instance, Figure 4.3 shows that the urban distribution spans a longer interval than the rural distribution, indicating the former has greater variation. Researchers are often interested in the measures of location and variation in the population as they constitute relatively clear summaries of important aspects of distributions. The numerical summaries of the population distribution are called *parameters*. Parameters are estimated using sample data.

4.3.1 Measuring Central Tendency

The three most common measures of location are the *mean*, the *median*, and the *mode*. The mode describes a typical measurement in terms of the most common outcome or most frequently occurring score. In Figure 4.2, the mode of each distribution is the dollar amount directly under the peak of the curve. A limitation in using the mode is that a distribution can have more than one. This indicates that the mode will not always have a unique value and, thus, cannot be recommended for general use.

In contrast to the mode, the *median* and *mean* are always unique values. The median is the middle-most score in a distribution. The median() function is used to find the median of the distribution. The best known and most frequently used measure of central tendency is the *mean*, or the average. The mean() function is used to find the mean of a distribution. The use of the median() and mean() functions is shown in Command Snippett 4.7.

Command Snippet 4.7: Syntax to obtain numerical summaries of location for per capita household expenditures.

```
## Marginal median household per capita expenditure
> median(household$Dollars)
[1] 159.8

## Marginal mean household per capita expenditure
> mean(household$Dollars)
[1] 212.5778
```

The median household per capita expenditure is $160, and the mean household per capita expenditure is $213. In symmetric distributions, the mean and median can be equal or nearly so. However, in asymmetric distributions, the two can differ, sometimes drastically so.

The mean and median computed previously summarize the marginal distribution, as Area is ignored. Though the marginal estimates are useful, the goal is to compute the conditional estimates of a typical household per capita expenditure for each area. Unfortunately, the mean() and median() functions, by themselves, do not allow for conditioning on Area. For conditional results, the tapply() function is used.

The `tapply()` function applies some other function (e.g., `mean()`) to a numeric response variable, conditional on the levels of a predictor variable, the latter being a *grouping* or *factor* variable. The arguments for the function are the quantitative vector (`X=`), the conditioning factor (`INDEX=`), and the function to be applied (`FUN=`). The `tapply()` function also takes optional arguments for the function being applied. For example, the argument `na.rm=TRUE`—which is an optional argument for `median()` and `mean()` for the treatment of missing values—can be appended as a fourth argument in the `tapply()` function.

The syntax for computing the mean household per capita expenditure conditioned on area is shown in Command Snippet 4.8. The output in that snippet shows that the mean for the urban area is more than twice that for the rural area. This is consistent with Figure 4.2 that shows the urban distribution being right-shifted relative to the rural distribution. This suggests that the average household per capita expenditure differs for urban and rural areas in the sample.

Command Snippet 4.8: Syntax to obtain numerical summaries of central tendency for per capita household expenditures conditioned on area.

```
> tapply(X = household$Dollars, INDEX = household$Area, FUN =
    mean)
Rural 157.4192
Urban 348.6887
```

4.3.2 Measuring Variation

When an analysis deals with at least two groups, as in the rural/urban comparisons, it is important to consider group differences in variability and well as location. Variability within the groups influences the evaluation of location differences. High within-group variability can be an overwhelming feature that can render location differences as irrelevant, or at least less relevant. On the other hand, low within-group variability can work to accentuate location differences.

Consider the examples of Figures 4.4 and 4.5. In both figures the mean difference between the distributions is the same. However, the large within-group variation in the rural distribution in Figure 4.4 makes the interpretation of group differences less clear for these data than for the data shown in Figure 4.5. In fact, it can be argued that the most important feature is the fact that the urban distribution is almost entirely contained within the rural distributions. This means, for example, that though the rural mean is lower than the urban mean, there are several rural households that are higher than the *urban* mean, and some that are higher than *any* urban households.

In contrast, consider Figure 4.5. There is essentially no overlap between the two distributions. This means that the mean difference also characterizes the difference between almost every pair of households from the two distributions. If we were to randomly select one rural and one urban household, the rural household would almost surely have a lower annual income. The same cannot be said of the overlapping distributions in Figure 4.4.

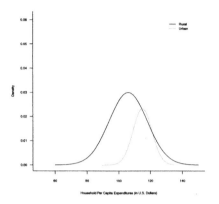
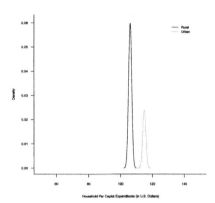

Figure 4.4: Simulated density plots for the distribution of household per capita expenditures (in dollars) conditioned on area showing large within-group variation.

Figure 4.5: Simulated density plots for the distribution of household per capita expenditures (in dollars) conditioned on area showing small within-group variation.

Two summary measures of variation—the standard deviation and variance—are based on the deviations of the data from the mean. The sd() and var() functions can be used to compute these quantities, respectively. Command Snippet 4.9 illustrates the use of the functions to find the variance and standard deviation for both the marginal and conditional distributions of household per capita expenditures.

Based on these conditional summaries, the rural households show less variation than the urban households. This is consistent with Figure 4.2 that shows the urban distribution being wider relative to the rural distribution. The average household per capita expenditure is more homogenous for rural than for urban households. There are some caveats regarding indexes of variation. Most notably, measures of variation are sensitive to asymmetry, and their values can be inflated by even a single extreme value. For this reason, the skewness of the distributions should be considered when comparing measures of variation computed on such distributions.

Another measure of variation that often gets reported in the educational and behavioral sciences, is the *standard error of the mean*. The idea underlying the standard error is that different samples drawn from the same population have different values of the sample mean. This is a consequence of random sampling and the fact that sample information is always incomplete relative to the population. The standard error of the mean is the standard deviation of all the possible sample means for a given sample size. As such, this measure offers an indication of the precision of the sample mean, when it is used as an estimate of the population mean. The smaller the

Command Snippet 4.9: Syntax to obtain numerical summaries of variation for per capita household expenditures.

```
## Marginal variance for the household per capita expenditures
> var(household$Dollars)
[1] 32221.84

## Marginal standard deviation for the household per capita
    expenditures
> sd(household$Dollars)
[1] 179.5044

## Variance for the household per capita expenditures
    conditioned on area
> tapply(X = household$Dollars, INDEX = household$Area, FUN =
    var)
Rural   9384.822
Urban 62564.272

## Standard deviation for the household per capita expenditures
    conditioned on area
> tapply(X = household$Dollars, INDEX = household$Area, FUN =
    sd)
Rural   96.8753
Urban 250.1285
```

standard error the greater the precision. The standard error for the mean is computed as

$$SE_{\bar{Y}} = \frac{SD_Y}{\sqrt{n}}. \tag{4.1}$$

where SD_Y is the standard deviation of the observed measurements on some variable Y. The standard error of the mean is computed for both the urban and rural households in Command Snippet 4.10. The standard error for the rural group is approximately four times smaller than that of the urban group ($\frac{6.01}{1.48} \approx 4$). This suggests that the sample mean for the rural households is a more precise estimate of the rural population mean than the sample urban mean is for the urban population. The use of the sample estimates and standard error for estimating population parameters is discussed further in Chapter 9.

4.3.3 Measuring Skewness

Skewness is a numerical measure that helps summarize a distribution's departure from symmetry about its mean. A completely symmetric distribution has a skewness

Command Snippet 4.10: Syntax to compute the standard error of the mean for both the rural and urban households.

```
## Standard deviations
> numerator <- tapply(X = household$Dollars, INDEX =
    household$Area, FUN = sd)

## Square root of sample sizes
> denominator <- sqrt(table(household$Area))

## Standard errors
> numerator / denominator
    Rural     Urban
1.482689 6.013678
attr(,"class")
[1] "table"
```

value of zero.[1] Positive values suggest a positively skewed (right-tailed) distribution with an asymmetric tail extending toward more positive values, whereas negative values suggest a negatively skewed (left-tailed) distribution with an asymmetric tail extending toward more negative values.

The **e1071** package[2] provides a function called skewness(), which computes the skewness value for a sample distribution based on three common algorithms. This function is supplied with the argument type=2 to compute *G1*, a slightly modified version of skewness that is a better population estimate (e.g., Joanes & Gill, 1998). Command Snippet 4.11 shows the use of skewness() for both the marginal and conditional distributions. To find the skewness for the conditional distributions, the argument type=2 provided in the skewness() function is appended as an additional argument in the tapply() function.

The output for the skewness() function suggests that both the urban and rural distributions are positively skewed, but more so for the rural group. The following guidelines are offered as help in interpreting the skewness statistic. Like all guidelines these should be used with a healthy amount of skepticism. All statistics should be interpreted in terms of the types and purposes of the data analysis, as well as the substantive area of the research.

- If $G_1 = 0$, the distribution is symmetric.

- If $|G_1| < 1$, the skewness of the distribution is slight.[3]

- If $1 < |G_1| < 2$, the the skewness of the distribution is moderate.

---

[1] Technically this is only true for an index of skewness that has been "corrected" or "standardized" so that the normal distribution has a skewness of zero. Skewness indices need not be zero for a normal distirbution in general.

[2] An alternative package is **moments**.

[3] $|G_1|$ indicates the absolute value of *G1* (cut off the sign).

Command Snippet 4.11: Functions to compute $G1$ for the marginal and conditional distributions of household per capita expenditures.

```
## Load e1071 package
> library(e1071)

## Skewness measure for the marginal distribution of household
     per capita expenditures
> skewness(household$Dollars, type = 2)
[1] 3.791977

## Skewness measure for the distribution of household per
     capita expenditures conditioned on area
> tapply(X = household$Dollars, INDEX = household$Area, FUN =
     skewness, type = 2)
    Rural      Urban
4.276803  2.728186
```

- If $|G_1| < 2$, the distribution is quite skewed.

The above guidelines indicate that both distributions in the example are severely positively skewed. Furthermore, the rural distribution is more asymmetric than the urban distribution. This is again consistent with Figure 4.2, which shows the rural distribution has a longer tail relative to its mean than the urban distribution. The distribution of rural households shows relatively less density for household per capita expenditures above the mean than below the mean. This asymmetry is even more evident for urban households.

4.3.4 Kurtosis

Kurtosis is often used as a numerical summarization of the "peakedness" of a distribution, referring to the relative concentration of scores in the center, tail, and shoulders. Normal distributions have a kurtosis value of zero and are called *mesokurtic*.[4] Distributions that reflect a more peaked and heavy-tailed distribution than the normal distribution have positive kurtosis values, and are said to be *leptokurtic*. Distributions which are flatter and lighter-tailed than the normal distribution have negative kurtosis values and are said to be *platykurtic*. Dyson and Cantab (1943, p. 360) suggest an "amusing mneumonic"—which was attributed to Gossett (Student, 1927)—for the above terms.

> Platykurtic curves, like the platypus, are squat with short tails. Leptokurtic curves are high with long tails, like the kangaroo—noted for "lepping".

Figures 4.6 and 4.7 depict distributions with different kurtosis values. The mesokurtic distribution is shown for a basis of comparison in both figures. The

---

[4]Again, technically this is only true for indices of kurtosis that have been "corrected" so that a normal distribution has a kurtosis of zero.

distributions in Figure 4.6 show positive kurtosis, whereas the distributions in Figure 4.7 show negative kurtosis.

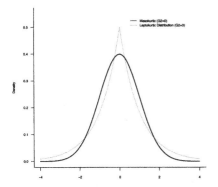 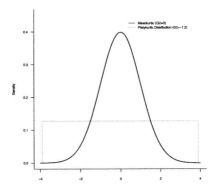

Figure 4.6: Kernel density estimate for a mesokurtic distribution (solid line) and a leptokurtic distribution (dotted line). The leptokurtic distributions are skinnier and more peaked than the mesokurtic distribution.

Figure 4.7: Kernel density estimate for a mesokurtic distribution (solid line) and a platykurtic distribution (dotted line). The platykurtic distribution is flatter than the mesokurtic distribution.

The `kurtosis()` function provided in the **e1071** package can be used to compute the sample kurtosis value for a distribution based on three common algorithms. We use this function with the argument `type=2` to compute $G2$, a slightly modified version of the kurtosis statistic that is a better population estimate of kurtosis (e.g., Joanes & Gill, 1998). Command Snippet 4.12 shows the use of `kurtosis()` to compute the kurtosis values for both the marginal and conditional distributions. To find the kurtosis for the conditional distributions, the argument `type=2` is provided in the `kurtosis()` function, again appended as an additional argument in the `tapply()` function.

The kurtosis statistics for the conditional distributions suggest that both distributions are severely leptokurtic indicating that these distributions are more peaked than a normal distribution. They also have more density in the tails of the distribution than we would expect to see in a normal distribution. One can see in Figure 4.2 that the rural distribution is even more peaked than the urban distribution.

While the kurtosis statistic is often examined and reported by educational and behavioral scientists who want to numerically describe their samples, it should be noted that "there seems to be no universal agreement about the meaning and interpretation of kurtosis" (Moors, 1986, p. 283). Most textbooks in the social sciences describe kurtosis in terms of peakedness and tail weight. Balanda and MacGillivray (1988, p. 116) define kurtosis as "the location- and scale free movement of probability mass

Command Snippet 4.12: Functions to compute $G2$ for the marginal and conditional distributions of household per capita expenditures.

```
## Kurtosis measure for the marginal distribution of household
   per capita expenditures
> kurtosis(household$Dollars, type = 2)
[1] 26.23688

## Kurtosis for the distribution of household per capita
   expenditures conditioned on area
> tapply(X = household$Dollars, INDEX = household$Area, FUN =
  kurtosis, type = 2)
   Rural    Urban
42.66108 14.02895
```

from the shoulders of a distribution into its center and tails ... peakedness and tail weight are best viewed as components of kurtosis." Other statisticians have suggested that it is a measure of the bimodality present in a distribution (e.g., Darlington, 1970; Finucan, 1964). Perhaps it is best defined by Mosteller and Tukey (1977) who suggest that like location, variation, and skewness, kurtosis should be viewed as a "vague concept" that can be formalized in a variety of ways.

4.4 SUMMARIZING THE FINDINGS

The APA manual (American Psychological Association, 2009) provides suggestions for presenting descriptive statistics for groups of individuals. The information should be presented in the text when there are three or fewer groups and in a table when there are more than three groups. While this number is not set in stone, we want to present results in a manner that will facilitate understanding. Typically we report measures of location, variation, and sample size for each group, at the very least. We present the results of our data analysis in box below.

4.4.1 Creating a Plot for Publication

In this section, a plot suitable for publication is constructed that addresses the research question regarding differences in living standards for urban and rural areas. Earlier in this chapter, there was an illustration of how to plot two density estimates on the same graph (see Command Snippet 4.5). For interpretability, it is helpful to add a legend to the plot or in some way label the distributions.

The plots of the two estimated densities are re-created for the rural and urban households using the syntax in Command Snippet 4.5. The text() function is used to add text to an existing plot. This function writes the text—provided as a quoted character string—given in the argument labels=. The text is drawn at a given coordinate in the plot specified using the arguments x= and y=. The argument pos=4 is also provided so that the text is written to the right of the given coordinate rather

Sample Write-Up

Empirical evidence on the process of urbanization has shown increased economic segregation among urban and rural households, as well as increased spatial differentiation of land uses (e.g., Leaf, 2002). The Socialist Republic of Vietnam, for the last decade, has experienced an industrialization characterized by economic growth and urbanization.

Statistical analysis shows that the typical household per capita expenditure is higher for urban households (M = $349, SE = $6) than for rural households (M = $157, SE = $1). The distribution for urban households (SD = $250) also shows more variation than the distribution for rural households (SD = $97) indicating that rural areas tend to be more homogeneous in their household per capita expenditures. This evidence is further strengthened by the difference in asymmetry, skewness $(G1)$ = 2.73 and skewness $(G1)$ = 4.28 for urban and rural households, respectively, and heavy-tailedness, kurtosis $(G2)$ = 42.66 and kurtosis $(G2)$ = 14.03, in the two sample distributions.

In contrast to their urban counterparts, the economic stimulation in rural areas of Vietnam seems not to have been as dynamic. The typical household for rural areas is only $15 U.S. above the poverty line. Furthermore, except for a rather small number of wealthier rural households, the majority of rural households show little variation in their household per capita expenditures. This shared level of poverty could be due to the fact that a substantial share of the populace living in rural areas of Vietnam are now unemployed or underemployed.

It is worth noting, that the poverty line—established in 1998 by the General Statistical Office at $119 U.S. (General Statistical Office, 2001)—is close to the mode of the rural expenditure per capita distribution, which could indicate that a small increase in household expenditure per capita is enough to shift many of the rural households to a position above the poverty line. This is one likely explanation for recent dramatic reductions in poverty rates in Vietnam. As the poverty line moves higher, further reductions in poverty rates are likely to be smaller in magnitude.

than centered. The full syntax is in Command Snippet 4.13, and the resulting graph is in Figure 4.8.

4.4.2   Using Color

In Figures 4.2 and 4.8, the two conditional distributions were differentiated by changing the line type. Another common way to differentiate between elements on a graph

Command Snippet 4.13: Syntax to plot the conditional densities on the same graph and differentiate the two by varying the line type.

```
> plot(d.rural, main = " ", xlab = "Household Per Capita
    Expenditures (in U. S. Dollars)", lty = "solid", bty = "l")
> lines(d.urban, lty = "dotted")
> text(x = 182, y = 0.0059, labels = "Urban Households", pos =
    4)
> text(x = 495, y = 0.0012, labels = "Rural Households", pos =
    4)
```

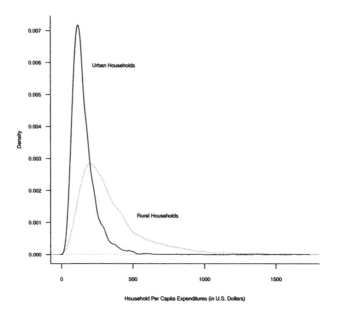

Figure 4.8: Plots of the kernel density estimates for the distribution of household per capita expenditures (in dollars) for the rural ($n = 4269$) and urban households ($n = 1730$).

is to use color. The col= argument can be added to either the plot() or lines() function. This argument takes a color which can be specified in many ways.

One way to specify color is to use the RGB color model.[5] This model superimposes varying intensities of red, green, and blue light to compose a broad palette of

---

[5]Most electronic devices, such as computer monitors use the RGB color space to manage colors, which makes it an ideal choice for viewing color which will be displayed on your screen.

colors. RGB colors are specified using the `rgb()` function. This function requires a numerical value for each of the three arguments, `red=`, `green=`, and `blue=`.

A value of 0 for any of the colors makes the associated color fully transparent. The maximum value for any color, which is set using the argument `maxColorValue=`, will make the color opaque. For the examples below, the maximum color value is set using `maxColorValue=255`.[6] Table 4.1 shows some common colors and their associated RGB values.

Table 4.1: Common Colors and Their Associated RGB Values Using a Maximum Color Value of 255

	RGB Value		
Color	Red	Green	Blue
White	0	0	0
Red	255	0	0
Green	0	255	0
Blue	0	0	255
Black	255	255	255

The RGB color values are supplied to the `col=` argument in the `plot()` function. To draw the density plot of the rural households in dark red, the `rgb` setting of `red=139`, `green=0`, `blue=0`, `maxColorValue=255` is used. The `rgb` combination of `red=0`, `green=0`, `blue=139`, `maxColorValue= 255` is supplied to the `col=` argument in the `lines()` function to specify that the density plot for the urban household be drawn in dark blue. Since colors are used to differentiate the two distributions, we also change the `lty=` argument in the `lines()` function to `"solid"`. Command Snippet 4.14 shows the syntax to create such a plot.

The `polygon()` function can be used to shade the densities being plotted. Similar to the `lines()` function, `polygon()` adds to an existing plot. We supply the assigned density as an argument to the `polygon()` function. We also use the optional argument `col=` to shade the densities in dark red and blue, respectively, by declaring them in the `rgb()` function. This time, however, an additional argument is added to the RGB color space. This fourth argument, `alpha=`, is called an *alpha* value, and it sets the degree of transparency of the color. The alpha value is set to 100 to make the color mostly transparent—again using the scale of 0 to 255, where 0 is completely transparent and 255 is opaque. Finally, the argument `lty=` is supplied to

---

[6]The maximum value of 255 is an artifact of restrictions in the hardware used in early personal computers which could encode 256 distinct numerical values (0–255) on a single 8-bit byte—the standard currently used in the computing industry. Higher-end digital processors typically use a 16-bit byte—or in some cases a 40-bit byte. Using a 16-bit byte increases the values which can be encoded on a single byte from 256 to 65,536. This allows 100 trillion colors to be represented rather than only 16.7 million using an 8-bit byte.

Command Snippet 4.14: Using the RGB color model in the `plot()` and `lines()` functions.

```
> plot(d.rural, main = " ", xlab = "Household Per Capita
    Expenditures (in U. S. Dollars)", bty = "l", col = rgb(red
    = 139, green = 0, blue = 0, maxColorValue = 255), lty =
    "solid")
> lines(d.urban, col = rgb(red = 0, green = 0, blue = 139,
    maxColorValue = 255), lty = "solid")
> text(x = 182, y = 0.0059, labels = "Urban Households", pos =
    4)
> text(x = 495, y = 0.0012, labels = "Rural Households", pos =
    4)
```

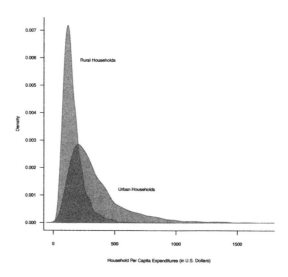

Figure 4.9: Kernel density plots for the distribution of household per capita expenditures (in dollars) conditioned on area. The density of household per capita expenditures for both distributions are shaded using the `polygon()` function. The plot appears in grayscale in this book, but should print in color on a computer.

the `polygon()` function to set the line type for the border of the density plot. The `type=` argument to the `plot()` function is used to control how data are plotted. The argument `type="n"` is used for no plotting, which can be useful to set up a plot for latter additions. The syntax is displayed in Command Snippet 4.15 and the resulting plot is displayed in Figure 4.9.

There are two additional features that often make plotting easier. First, one common way to specify color is to use a character string that provides a named

Command Snippet 4.15: Using the RGB color space to semitransparently shade a density plot in the polygon() function.

```
> plot(d.rural, main = " ", xlab = "Household Per Capita
    Expenditures (in U. S. Dollars)", bty = "l", type = "n")
> polygon(d.rural, col =  rgb(red = 139, green = 0, blue = 0,
    alpha = 100, maxColorValue = 255), lty = "solid")
> polygon(d.urban, col= rgb(red = 0, green = 0, blue = 139,
    alpha = 100, maxColorValue = 255), lty = "solid")
> text(x = 182, y = 0.0059, labels = "Urban Households", pos =
    4)
> text(x = 495, y = 0.0012, labels = "Rural Households", pos =
    4)
```

color. For example, col="red" is the same as col=rgb(red=255, green=0, blue=0, maxColorValue=255). R recognizes 657 color names, and a list of these names along with their colors is provided at http://www.stat.columbia.edu/~tzheng/files/Rcolor.pdf. These names can also be accessed and printed to the R terminal window by using the colors() function with no arguments. It should be noted, however, that use of the named colors does not allow for transparency through an alpha value.

Secondly, an alternative to providing the specific *x*- and *y*-coordinates in the text() function, is to use the command locator(1). This allows for a point-and-click specification of the location of the text in the graph. The syntax for using this functionality is shown in Command Snippet 4.16. The argument pos=4 will position the text to the right of the clicked location, rather than centering the text at that location.

Command Snippet 4.16: Syntax to use the locator() function to position the text on a graph.

```
> text(locator(1), labels = "Rural Households", pos = 4)
```

4.4.3   Selecting a Color Palette

Choosing a good color palette is important. Physiological studies have suggested that color can influence a reader's attention, their perception of area, and their ability to distinguish different groups (see Cleveland & McGill, 1983; Ihaka, 2003, for more details). Unfortunately, choosing a color palette is also a difficult exercise since humans can distinguish millions of distinct colors and the sheer size of this "search space" makes it hard to find good color combinations. This problem is compounded, when colors are evaluated in the presence of other colors, since the appearance of colors change when they are displayed with other colors. Some guidance for choosing a color palette can be found in the principles of design that have been used by artists for centuries (see Brewer, 1999; Zeileis & Hornik, 2006; Zeileis, Hornik, & Murrell, 2007). One package that uses these principles is the **RColorBrewer** package.

The **RColorBrewer** package can be used for selecting color palettes for plots. It provides a rich collection of suitable palettes for coding various types of information. There are three types of color schemes that are distinguished: (a) qualitative, (b) sequential, and (c) diverging (Brewer, 1999; Brewer, Hatchard, & Harrower, 2003; Harrower & Brewer, 2003). The first is used for coding categorical information and the latter two are for coding ordinal and continuous variables. Each color scheme has several different palettes to choose from. Brewer et al. (2003) presents the palette choices available within each color scheme.

The function brewer.pal() in the **RColorBrewer** can be used to select a suitable color palette for the density plot. This function takes two arguments. The first, n=, takes a value indicating the number of colors being used in the plot. The second, name=, takes a character string indicating the name of the color palette.

Returning to the example, color is being used to differentiate the urban and rural households, so a palette from the qualitative color scheme is appropriate for the plot. The palette Set1 is one such color palette. Command Snippet 4.17 shows the syntax to determine colors for the plot of household per capita expenditures conditioned on area.

Command Snippet 4.17: Choosing a qualitative color palette for two groups.

```
## Load RColorBrewer package
> library(RColorBrewer)

## Choosing a color palette for two groups
> brewer.pal(n = 2, name = "Set1")
[1] "#E41A1C" "#377EB8" "#4DAF4A"
Warning message:
In brewer.pal(n = 2, name = "Set1") :
  minimal value for n is 3, returning requested palette with 3
      different levels

## Load the colorspace package
> library(colorspace)

## Convert the hexadecimal values to RGB
> hex2RGB(brewer.pal(n = 3, name = "Set1"))
              R          G          B
[1,] 0.79240771 0.01512116 0.01683345
[2,] 0.05014993 0.23774340 0.50961270
[3,] 0.09217427 0.46004311 0.08562281
```

In Command Snippet 4.17, a warning is printed since the minimum number of levels for the palette is three, and the colors associated with the three-color palette are printed. For two groups, the first two values can be used. The color values returned are expressed as hexadecimal values. These values can be issued directly to the col= arguments in the plot(), lines(), and polygon() functions. Alternatively, the hex2RGB() function from the **colorpace** package can be used to convert these values to RGB values. Converting them to RGB values allows the use of an alpha value to make the colors semitransparent.

Command Snippet 4.17 shows the syntax for converting the hexadecimal color values to RGB values. Note that the RGB values are on a scale from 0 to 1 rather than 0 to 255. To use these values the maxColorValue= argument in the rgb() function would need to be set to 1.

When choosing a color palette, it is important to consider whether or not the work will be published in color. Many colors do not translate well to black and white, so the information that is conveyed in the plot via color may be lost. If information on a plot is best displayed using color, but will be published in black and white, a good option is to use a grayscale palette. Another consideration is readers who are color-blind. Given that about 10% of men have some degree of red–green color blindness, it is worth trying to avoid graphics that communicate information primarily through red-green distinctions (Lumley, 2006). The dichromat() function in the **dichromat** package can be used to approximate the effect of the two common forms of red–green color blindness, protanopia and deuteranopia, on the palette you have chosen.

4.5 EXTENSION: ROBUST ESTIMATION

The sample mean, variance, and standard deviation can be inordinately influenced by outliers that may be present in the sample data. The outliers usually consist of a small proportion of extreme observations in one or the other tail of the distribution. Because of this, in some distributions—such as skewed distributions—the sample mean and variance are not good representatives of the typical score and variation in the population. Thus, the examination of these statistics may offer a poor summary of how the populations differ and also of the magnitude of those differences. Better estimates of the typical score and variation in such cases are computed using robust estimates. Robust estimates reduce the effects of the tails of a sample distribution and outliers by either trimming or recoding the distribution before the estimates are computed. An advantage of a robust estimate is that its associated standard error will typically be smaller than its conventional counterpart.

4.5.1 Robust Estimate of Location: The Trimmed Mean

One strategy for reducing the effects of the tails of a distribution is simply to remove them. This is the strategy employed by trimming. To find a trimmed mean, a certain percentage of the largest and smallest scores are deleted and the mean is computed using the remaining scores. Table 4.2 shows an example of 20% trimming on a small data set. The original data in the first column contains the outlying case of 250. The outlier results in a mean value of 40, which is well out of the range of the majority of the scores—it does not summarize a typical score in the distribution. Based on Table 4.2, the resulting trimmed mean is

$$\frac{13 + 14 + 15 + 19 + 22 + 22}{5} = 21.$$

Table 4.2: Small Data Set (original data) Showing Results of Trimming (trimmed data) and Winsorizing (winsorized data) Distribution by 20%

Original Data	Trimmed Data	Winsorized Data
10		13
12		13
13	13	13
14	14	14
15	15	15
19	19	19
22	22	22
22	22	22
23		22
250		22
$M = 40.00$	$M = 17.50$	$M = 17.50$
Var $= 5465.78$	Var $= 16.30$	Var $= 18.06$

Compare this value to the conventional sample mean of 40 for the original data in Table 4.2. The conventional mean is highly influenced by the sample observation of 250. The trimmed mean is robust in the sense that the single extreme score of 250 in the tail does not exert undue influence and the trimmed mean is more similar to the majority of the nonoutlying scores. The optional argument `tr=` in the `mean()` function can be used to compute a trimmed mean. This argument can also be added as an additional argument in the `tapply()` function. Command Snippet 4.18 shows the syntax to compute a 20%, or .20, trimmed mean for the marginal and conditional distributions of household per capita expenditures.

One last note that bears mentioning is that the median is also sometimes used as a robust estimate of location. In fact, the median is just a trimmed mean with the percentage of trim equal to

$$\frac{1}{2} - \frac{1}{2n}, \tag{4.2}$$

where n is the sample size. This is an excessive amount of trimming. For example, based on the data in Table 4.2, the trimming would be

$$0.50 - (0.05) = .45.$$

The median trims 45% of the data!

Command Snippet 4.18: Syntax to obtain robust estimates of the central tendency for the per capita household expenditures.

```
## Marginal 20% trimmed mean
> mean(household$Dollars, tr = 0.2)
[1] 168.9201

## Conditional 20% trimmed means
> tapply(X = household$Dollars, INDEX = household$Area, FUN =
    mean, tr = 0.2)
Rural 139.4673
Urban 291.8670

## Marginal median
> median(household$Dollars)
[1] 159.8

## Conditional medians
> tapply(X = household$Dollars, INDEX = household$Area, FUN =
    median)
Rural 134.870
Urban 278.505
```

4.5.2 Robust Estimate of Variation: The Winsorized Variance

It is also possible to compute a robust estimate for the variation in a data set. A robust estimate of variation can be obtained by recoding extreme observations to be less extreme. This recoding is known as Winsorizing.[7] In essence, Winsorizing the distribution recodes a particular percentage of scores in the upper tail of the distribution to the next smallest score. Likewise, a certain percentage of scores in the lower tail of the distrbution are recoded to the next largest score. Table 4.1 shows the result of Winsorizing a distribution by 20%. The variance of the original data in the first column is highly influence by the outlier of 250. Its value is 5465.78, which appears to grossly overrepresent the variability among the majority of scores.

The Winsorized variance is computed as

$$\hat{\sigma}_W^2 = \frac{1}{n-1} \sum (w_i - \hat{\mu}_W)^2, \tag{4.3}$$

where w_i are the values in the Winsorized distribution, and $\hat{\mu}_W$ is the Winsorized mean. The Winsorized standard deviation can be found by taking the square root of the Winsorized variance. Computing a Winsorized variance for the data in Table 4.2 results in 18.06. Compare this value to the conventional variance of 5465.78.

---

[7]Among the mathematicians recruited by Churchill during the Second World War was one Charles Winsor. For his efforts in removing the effects of defective bombs from the measurement of bombing accuracy, he received a knighthood and we received a new statistical tool—Winsorized means.

The Winsorized variance is used as an accompanying measure of variation for the trimmed mean. The `winvar()` function, available in the package **WRS**, can be used to compute this robust measure of variability.[8] The argument `tr=` sets the percentage of Winsorizing. Command Snippet 4.19 provides the syntax for loading the **WRS** package. It also provides the syntax for computing the 20% Winsorized variance for both the marginal and conditional distributions.

Command Snippet 4.19: Syntax to obtain the Winsorized variance.

```
## Load the WRS package
> library(WRS)

## Marginal 20% Winsorized variance
> winvar(household$Dollars, tr = 0.2)
[1] 4640.54

## Conditional 20% Winsorized variances
> tapply(X = household$Dollars, INDEX = household$Area, FUN =
    winvar, tr = 0.2)
Rural    1820.721
Urban 12831.518

## Conditional 20% Winsorized standard deviation for the rural
    households
> sqrt(tapply(X = household$Dollars, INDEX = household$Area,
    FUN = winvar, tr = 0.2))
Rural      Urban
 42.66990 113.27629
```

Based on the output from Command Snippets 4.18 and 4.19, these robust estimators tell a slightly different story than the conventional estimates. Although the conditional means suggest urban and rural differences in household per capita expenditures, the 20% trimmed means suggest that the economic difference between a typical urban household and a typical rural household is less pronounced. The 20% Winsorized variances, while still suggesting a great deal of variation within each area, also suggest that much of that variation is likely due to the influence of outlying households. This can be seen in Table 4.3, in which both the conventional and robust estimates are provided for comparison.

Decisions about when to use a robust estimator such as a trimmed mean or Winsorized variance and also about how much to trim or Winsorize are not trivial tasks. Trimming or Winsorizing 20% of the data is common, while for very heavy tailed distributions, 25% may be more appropriate (Rosenberger & Gasko, 1983; Wilcox, 2005). Although statistically sound, many disciplines within the behavioral and educational sciences do not report or perform analyses with robust estimates. If an applied researcher in these fields chooses to use such estimates, she/he may need

---

[8]The **WRS** package needs to be installed from R-Forge (see Chapter 1).

Table 4.3: Comparison of Conventional and Robust Estimates of Location and Variation for Conditional Distributions of Household per Capita Expenditures

| | Rural Households | | Urban Households | |
Estimates	*M*	SD	*M*	SD
Conventional	157	97	349	250
Robust	139	43	292	113

to defend that choice or at the very least offer references that do. The following is an example of the write-up that may accompany the use of robust estimates.

> **Sample Write-Up**
>
> Because of the asymmetry of the observed distributions, the means reported were trimmed 20% and the standard deviations Winsorized by 20%; see Wilcox (2004) for details of these procedures. Both the graphical and statistical evidence indicate the typical household per capita expenditure is higher for urban households ($M = \$292$) than for rural households ($M = \$139$). The distribution for urban households (SD = 43) also shows more variation than the distribution for rural households (SD = 113) indicating that rural areas tend to be more homogeneous in their household per capita expenditures.

4.6 FURTHER READING

Pearson (1894) and Pearson (1895) are of historical importance for their introduction of statistical terms and symbols (e.g., skewness, kurtosis, σ) and more narrative details of these origins are available in Fiori and Zenga (2009) and David (1995). An introduction to the ideas underlying robust estimates can be found in Huber (1981), Hampel, Ronchetti, Rousseeuw, and Stahel (1986), Staudte and Sheather (1990), and Wilcox (2001, 2005). Other views can be found in Andrews et al. (1972), Rosenberger and Gasko (1983), and Dixon and Yuen (1999). Excellent views on designing graphs can be found in Tufte (1990) and Wainer (1984). Research on the perception of color can be found in Zhang and Montag (2006) and Kaiser and Boynton (1996). Research on the choices of color palettes in graphs are provided in Zeileis and Hornik (2006) and Zeileis et al. (2007).

PROBLEMS

4.1 Data on U.S. citizens were collected in 2000 by the U.S. Census Bureau. The data set *LatinoEd.csv*—a subset of the Census data—contains data for a sample of

Latino immigrants who were naturalized U.S. citizens living in Los Angeles. The variable Achieve in this data set provides a measure of educational achievement (see the codebook for more information regarding the interpretation of this variable).

a) Construct a plot of the kernel density estimate for the marginal distribution of the educational achievement variable. Discuss all interesting features of this plot.

b) Examine plots of the variable Achieve conditioned on the variable English to compare the educational achievement of Latino immigrants who are fluent in English and those who are not. Create a single publishable display that you believe is the best visual representation of the results of this analysis. In constructing this display, think about the substantive points you want to make and create a graph that best allows you to highlight these conclusions. Write a brief paragraph explaining why you chose to construct your graph the way you did and how it helps answer the research question of interest.

c) Compute appropriate numerical summaries of achievement conditioned on English fluency. Use these summaries (along with evidence culled from your plot) to provide a full *comparison of the distributions* of achievement scores for immigrants who are fluent in English and those immigrants who are not fluent in English. Be sure to make comparisons between the measures of center and variation in the distributions. Use the problem context to help you write your answer. Be selective in what you report, remembering that you want to be succinct yet thorough.

4.2   Hearing of your more sophisticated data analysis, the Center for Immigration Studies has asked you to write *no more than a one-page* editorial for your local newspaper summarizing your results. In particular, they would like you to address the following:

- Are there achievement disparities between immigrants who are fluent in English and immigrants who are not fluent in English? How do the achievement scores for these groups compare to one another?

- How much variation is there in the achievement scores of immigrants who are fluent in English? If so, how does this compare to the variation in the achievement scores of their nonfluent counterparts?

- Does your analysis tell us anything about how schools could improve the achievement scores of their nonfluent students? If you could use the bully pulpit of your local newspaper to bring pressure on the education system to do additional research on improving these achievement scores, what type of study, or what type of evidence might you encourage them to pursue?

As you write this editorial, remember that in general, newspaper readers are not familiar with technical terms, so be sure to express your statistical evidence in a way that would be understandable to the general public.

CHAPTER 5

EXPLORATION OF MULTIVARIATE DATA: COMPARING MANY GROUPS

Researchers from nearly every social and physical science discipline have found themselves in the position of simultaneously evaluating many questions, testing many hypothesis, or comparing many point estimates.

—A. Gelman, J. Hill, & M. Yajima (in press)

In Chapter 3, the Vietnam Living Standards Survey (VLSS) was introduced to answer three research questions to inform an evaluation of policies and programs in Vietnam. These research question were:

1. What is the age distribution for the Vietnamese population?

2. Are there differences in the annual household per capita expenditures between the rural and urban populations in Vietnam?

3. Are there differences in the annual household per capita expenditures between the seven Vietnamese regions?

The first question regarding the nature of the age distribution was addressed in Chapter 3. The second question concerning whether the living standards for urban

Comparing Groups: Randomization and Bootstrap Methods Using R
First Edition. By Andrew S. Zieffler, Jeffrey R. Harring, & Jeffrey D. Long
Copyright © 2011 John Wiley & Sons, Inc.

and rural Vietnamese is equitable was addressed in Chapter 4. In this chapter the third research question of whether there are economic differences by region of the country is addressed.

5.1 GRAPHING MANY CONDITIONAL DISTRIBUTIONS

To address the research question, the VLSSperCapita.csv data again needs to be read into R using the read.table() function. As always, the data frame should be examined to be sure the data were read in correctly and to evaluate whether there are problems that may exist in the data. Command Snippet 5.1 provides an example of this syntax. The marginal distribution of per capita household expenditures for all 5999 households is again examined. The output and results of these commands are not presented here, nor are the interpretations, since they are the same as was presented in Chapter 4.

Command Snippet 5.1: Code to read in the per capita data, examine the data frame, and plot the marginal distribution of household per capita expenditures.

```
## Read in the data
> household <- read.table("/Documents/Data/VLSSperCapita.csv",
      header = TRUE, sep = ",", row.names = "ID")

## Examine the data frame
> head(household)
> tail(household)
> str(household)
> summary(household)

## Plot the marginal distribution
> plot(density(household$Dollars), xlab = "Dollars", main = "
      ")
```

Examining the marginal distribution is useful in an initial examination of the data, but it does not help in answering the research question about regional differences. In order to address questions about regional differences, conditioning on the Region variable is necessary. In Chapter 4, indexing was introduced as a way to access subsets of a vector or data frame. Examination of the codebook for these data provides the region names associated with each of the integer values in the Region variable. Command Snippet 5.2 shows the use of indexing to create subsets of the household per capita expenditures for each region.

The plot() and lines() functions can then be used in conjunction with the density() function to produce the conditional density plots. Recall from Chapter 4 that the first region's density is plotted using the plot() function, and the remaining regions' densities are plotted using the lines() function. Since there are several groups' densities that need to be plotted, differentiation of the regions is best performed through color rather than line type. Command Snippet 5.3 shows

Command Snippet 5.2: The use of indexing to create subsets of the household per capita expenditures for each region.

```
> central.coast <- household$Dollars[household$Region == 1]
> central.highlands <- household$Dollars[household$Region == 2]
> mekong.delta <- household$Dollars[household$Region == 3]
> north.coast <- household$Dollars[household$Region == 4]
> northern.uplands <- household$Dollars[household$Region == 5]
> red.river.delta <- household$Dollars[household$Region == 6]
> south.east <- household$Dollars[household$Region == 7]
```

the syntax to use the **RColorBrewer** package (see the previous chapter) to select a qualitative color palette, plot the household per capita expenditures densities for all seven regions, and add a legend. The resulting graph is shown in Figure 5.1.

Command Snippet 5.3: Plotting the densities of the household per capita expenditures for all seven regions.

```
## Determine colors for the seven regions
> library(RColorBrewer)
> brewer.pal(n = 7, name ="Set1")
[1] "#E41A1C" "#377EB8" "#4DAF4A" "#984EA3" "#FF7F00"
    "#FFFF33" "#A65628"

## Plot the density for the first region
> plot(density(central.coast), main = " ", xlab = "Household
    Per Capita Expenditures (in U.S. Dollars)", bty = "l",
    xlim = c(0, 3100), ylim = c(0, 0.008), col = "#E41A1C")

## Add the densities for the remaining regions
lines(density(central.highlands), col = "#377EB8")
lines(density(mekong.delta), col = "#4DAF4A")
lines(density(north.coast), col = "#984EA3")
lines(density(northern.uplands), col = "#FF7F00")
lines(density(red.river.delta), col = "#FFFF33")
lines(density(south.east), col = "#A65628")

## Add a legend
legend(x = 2000, y = 0.0075, legend = c("Central Coast",
    "Central Highlands", "Mekong Delta", "North Coast",
    "Northern Uplands", "Red River Delta", "South East"),
    lty="solid", col = c("#E41A1C", "#377EB8", "#4DAF4A",
    "#984EA3", "#FF7F00", "#FFFF33", "#A65628"))
```

Examining the plot produced in Figure 5.1, it is very difficult to interpret the differences in per capita household expenditures between the seven regions. The primary difficulty is that having many superimposed densities on the same plot, especially when those distributions overlap, makes it very difficult to make out group differences. There are two solutions to this difficulty, panel plots and side-by-side box-and-whiskers plots.

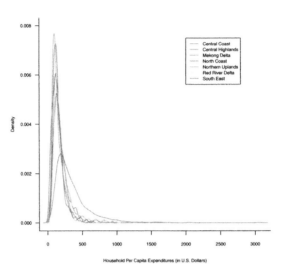

Figure 5.1: Kernel density plots for the distribution of household per capita expenditures (in dollars) conditioned on region. The graph is problematic as the various regions are difficult to sort out. The plot appears in grayscale in this book, but should print in color on a computer.

5.1.1   Panel Plots

One solution to the difficulty of many superimposed densities is to plot the kernel density estimate for each region in a different graph or *panel*. All the panels have the same scale, so that all the graphs are comparable.

To permit paneling, the graphical parameters in R need to be changed. The graphical parameter mfrow= can be used to draw multiple panels on the graphics device or window. This argument takes a vector of the form c(*rows, columns*), where *rows* and *columns* are values to determine the array of panels in the graphical device. Figures 5.2 and 5.3 show two different panel arrays.

Graphical parameters were encountered in past chapters, but were not explicitly named as such. Arguments such as lty=, bty=, and col= are all examples of graphical parameters. These parameters all have default values in R, but the values can be changed to produce certain graphical effects, such as dotted lines, colors, etc. By referring to these parameters in functions such as plot() or lines() the parameters are changed *locally*, meaning that they only change for a particular graph. The par() function changes graphical parameters *globally*, for every plot created

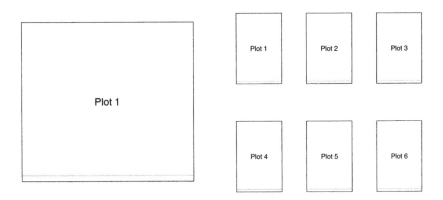

Figure 5.2: The default panel array is mfrow=c(1,1). This array has one row and one column.

Figure 5.3: This panel array is mfrow=c(2,3). This array has two rows and three columns.

until the R session is ended or the parameters are reset. It is always a good idea to return to the default settings after the global parameters have been altered for construction of a specific graph.

Command Snippet 5.4 shows the syntax to display the density estimates for each region on a different panel. First, par(mfrow=c(4,2)) is used to display the plots using four rows and two columns. To draw on a different panel, the plot() function is used. To label each panel, the argument main= is utilized in the plot() function. The x- and y-limits are also set to identical values in each of the seven panels. Furthermore, the x-label needs to be written for each plot. Lastly, the par() function is used again to restore the default parameter of one panel being displayed in the graphing device. The resulting panel plot is shown in Figure 5.4. The panel plots in Figure 5.4 clearly show the density for each region. Forcing the same axis scales facilitates comparison. For example, it is clear that the peak of the South East region is the lowest of all the regions.

5.1.2   Side-by-Side Box-and-Whiskers Plots

Another method of displaying information about several distributions simultaneously, is side-by-side box-and-whiskers plots. To produce side-by-side box-and-whiskers plot, the boxplot() function is used. The syntax details of this function are provided in Chapter 4. Command Snippet 5.5 presents the syntax for producing the side-by-side box-and-whiskers plot displayed in Figure 5.5. Figure 5.5 shows that all the distributions are positively skewed with several potential outliers. There are median differences among the regions with the South East having the largest value. The South East distribution is also the most varied for the innermost 50%, as its box length is greater than the other regions. The South East also has the most extreme outlier, with at least one household per capita value greater than $3000 (households with identical values are superimposed).

Command Snippet 5.4: Syntax for panel plots of the per capita household expenditures data conditioned on region.

```
## Set plotting array
> par(mfrow=c(4, 2))

## Define parameter options for the plots
> my.bty <- "l"
> my.xlab <- "Household Per Capita Expenditures (in U.S.
    Dollars)"
> my.xlim <- c(0, 3100)
> my.ylim <- c(0, 0.008)

## Panel 1
> plot(density(central.coast), main = "Central Coast", xlab =
    my.xlab,  bty = my.bty, xlim = my.xlim, ylim = my.ylim)

## Panel 2
> plot(density(central.highlands), main = "Central Highlands",
    xlab = my.xlab,  bty = my.bty, xlim = my.xlim, ylim =
    my.ylim)

## Panel 3
> plot(density(mekong.delta), main = "Mekong Delta", xlab =
    my.xlab,  bty = my.bty, xlim = my.xlim, ylim = my.ylim)

## Panel 4
> plot(density(north.coast), main = "North Coast", xlab =
    my.xlab,  bty = my.bty, xlim = my.xlim, ylim = my.ylim)

## Panel 5
> plot(density(northern.uplands), main = "Northern Uplands",
    xlab = my.xlab,  bty = my.bty, xlim = my.xlim, ylim =
    my.ylim)

## Panel 6
> plot(density(red.river.delta), main = "Red River Delta", xlab
    = my.xlab,  bty = my.bty, xlim = my.xlim, ylim = my.ylim)

## Panel 7
> plot(density(south.east), main = "South East", xlab =
    my.xlab,  bty = my.bty, xlim = my.xlim, ylim = my.ylim)

## Reset plotting array to default
> par(mfrow = c(1, 1))
```

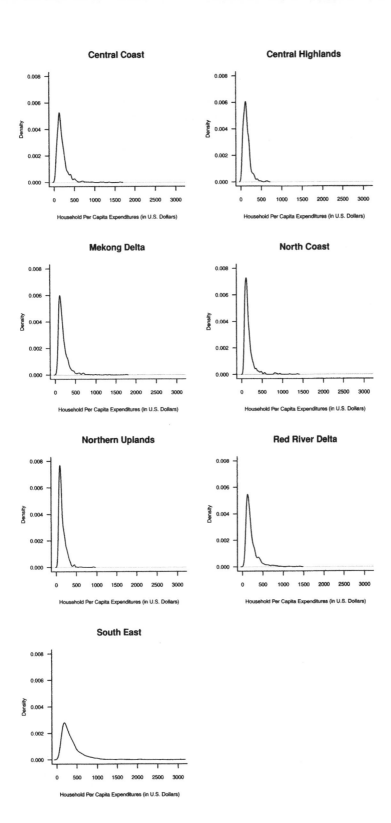

Figure 5.4: Panel plot of the density of household per capita expenditures (in dollars) conditioned on region.

Command Snippet 5.5: Syntax to produce side-byside box-and-whiskers plots of the household per capita expenditures (in dollars) conditioned on region.

```
> boxplot(central.coast, central.highlands, mekong.delta,
    north.coast, northern.uplands, red.river.delta, south.east,
    names = c("Central Coast", "Central Highlands", "Mekong
    Delta", "North Coast", "Northern Uplands", "Red River
    Delta", "South East"))
```

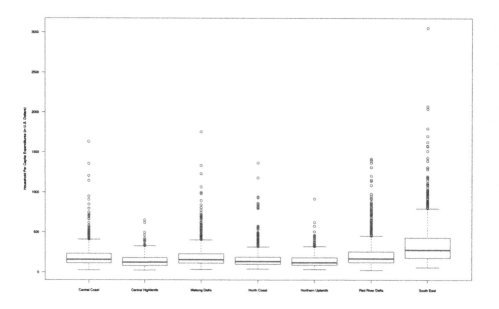

Figure 5.5: Side-by-side box-and-whiskers plots of household per capita expenditures (in U.S. dollars) conditioned on region.

5.2 NUMERICALLY SUMMARIZING THE DATA

The density plots in Figure 5.4 and the side-by-side box-and-whiskers plots in Figure 5.5 indicate that each distribution is positively skewed with potential outliers. Because of the extreme skewness and potential outliers in these distributions, we are inclined to use the more robust estimates for mean and variance. Recall that robust estimates are less affected by extreme scores than conventional estimates and, thus, are arguably more desirable summaries. The focus here is on the trimmed mean and Winsorized variance introduced in the last chapter. Since there are multiple distributions, it is advantageous to use some of the programming knowledge developed in the past chapters. For instance, rather than taking the square root of each of the seven Winsorized variances, we will chain the `tapply()` function to the `sqrt()` function. The syntax is provided in Command Snippet 5.6, along with syntax to obtain the sample sizes for each region.

Command Snippet 5.6: Robust estimates for the parameters for household per capita expenditures conditioned on region.

```
## Obtain the conditional 20% trimmed means
> tapply(X = household$Dollars, INDEX = household$Region, FUN =
    mean, tr = 0.2)
        1        2        3        4        5        6
164.7023 126.1242 162.4833 137.9229 126.7494 176.9796
        7
289.1686

## Load the WRS package
> library(WRS)

## Obtain the conditional 20% Winsorized standard deviations
> sqrt(tapply(X = household$Dollars, INDEX = household$Region,
    FUN = winvar, tr = 0.2))
        1        2        3        4        5
 59.27969 44.81304 57.21359 42.48159 45.73762
        6        7
 66.34581 119.69151

## Obtain the sample size for each region
> table(household$Region)

   1    2    3    4    5    6    7
 754  368 1112  708  859 1175 1023
```

The numerical evidence shows there are regional differences in both the typical household per capita expenditure and in the variation within regions. Consistent with the plots, the trimmed mean for the South East region is the largest, and so is its Winsorized variance. It is yet to be seen whether the sample differences suggests differences in the population.

5.3 SUMMARIZING THE FINDINGS

When the information for several groups is presented, the statistical data should be summarized in either a table or graph, rather than in the text. This makes it easier for readers to take in and understand the information. In the narrative describing the results, reference is made to the table(s) and/or graph(s). The narrative of the manuscript is used to emphasize particular analyses that highlight interpretations, rather than to re-report all of the statistics in the text (American Psychological Association, 2009).

Regarding the results of the example, there are several issues that should be considered. Because of the asymmetry and outliers in the data (see Figure 5.5), it is desirable to report the values for the trimmed mean and Winsorized standard deviations, rather than the conventional estimates. The sample size for each region is also reported to further inform comparisons. When reporting numerical values—both in text and tables, the APA manual recommends that they be "express[ed] to the number of decimal places that the precision of measurement justifies, and if possible, carry all comparable values to the same number of decimal places" (American Psychological Association, 2009, p. 137). For the example, rounding to the nearest U.S. dollar seems appropriate given that the original responses were recorded to the nearest dong, which is the standard Vietnamese currency. The decimal points in the data result from the conversion from Vietnamese dong to U.S. dollars.

5.3.1 Writing Up the Results for Publication

For the write-up, we will use substantive content knowledge to help summarize and interpret the statistical findings.

Sample Write-Up

The graphical and statistical evidence shows there are sample differences in the household per capita expenditures between the regions. The numerical summaries obtained from the analysis are reported in Table 5.2. The South East region has the highest household typical per capita expenditure. The distribution of expenditures per capita in the South East region, which includes Ho Chi Minh City and the hinterland, is typically higher and more variable, compared with other regions. Also, more than 75% of the people sampled from the South East region are above the poverty line of $119.32. The distributions in the other six regions are fairly similar. It is worthwhile to note that in the two most impoverished regions, the Central Highlands and Northern Uplands, close to 50% of the households are below the poverty line.

5.3.2 Enhancing a Plot with a Line

Figure 5.4 can be used in a publication, but to make the plot more informative, the poverty line could be added to each panel. As discussed in Chapter 4, a line can be added to a plot using the abline() function. This draws a line to an existing plot or panel. To add the poverty line to each panel in Figure 5.4, the abline() function needs to be executed after each of the plot() calls. Command Snippet 5.7 shows the syntax for adding the poverty line to the figure. The syntax produces the paneled graph in Figure 5.6. As the graph shows, the distributions of most regions have typical scores at or very close to the poverty line. The South East region appears to be the only one whose curve peak is slightly to the right of the poverty line (i.e., higher).

Table 5.1: Mean, Standard Deviation and Sample Sizes for Household per Capita Expenditures (in dollars) Conditioned on Region[a]

Region	M_t	SD_W	n
Central Coast	165	59	754
Central Highlands	126	45	368
Mekong Delta	162	57	1112
North Coast	138	42	708
Northern Uplands	127	46	859
Red River Delta	177	66	1175
South East	289	120	1023

[a]The mean was trimmed by 20%. The standard deviation was computed from a variance that was Winsorized by 20%.

Command Snippet 5.7: Syntax to add the poverty line to each panel of a plot. This syntax needs to be issued after each plot() call.

```
> abline(v = 119, lty = "dashed")
```

5.4 EXAMINING DISTRIBUTIONS CONDITIONAL ON MULTIPLE VARIABLES

Now that regional differences in household per capita expenditures have been identified in the sample—and in Chapter 4 urban and rural differences were evident—there is interest in examining urban and rural differences within each region.

Indexing can again be used to create objects that contain the household per capita expenditures for a particular area (urban or rural) for each region. As discussed in Chapter 2, the logical operator & can be used to subset a particular region *and* area. Command Snippet 5.8 shows the syntax for creating these subsets of data for one region. The syntax needs to be repeated for all 7 regions (i.e., a total of 14 objects).

A side-by-side box-and-whiskers plot can be created in the same manner as presented in Command Snippet 5.6. The help menu for the boxplot() function offers additional arguments to fine-tune the plot. The panel plot for the density estimates can also be constructed. This is carried out in a similar manner as the previous panel plot (see Command Snippet 5.5). This time, the lines() function is used to add an additional density estimate to the same panel. Command Snippet 5.9 shows the

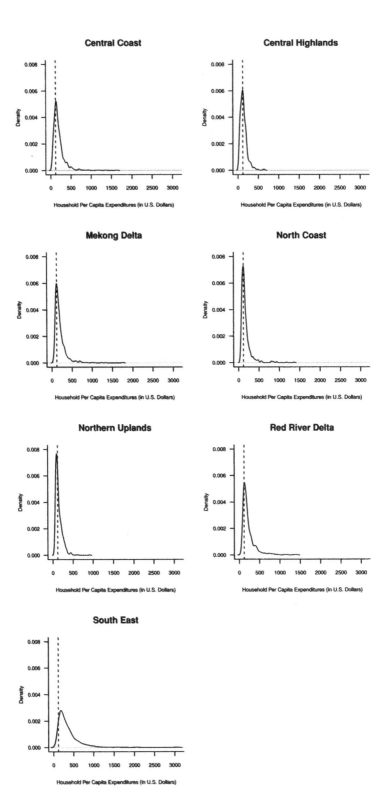

Figure 5.6: Plot of the estimated density of household per capita expenditures (in U.S. dollars) conditioned on region with the poverty line (shown as a dashed line) demarcated.

syntax used in each of the seven panels. The resulting panel plot for all seven regions is displayed in Figure 5.7.

Command Snippet 5.8: The use of indexing to create subsets of the household per capita expenditures for the Central Coast region.

```
> central.coast.urban <- household$Dollars[household$Region ==
    1 & household$Area == "Urban"]
> central.coast.rural <- household$Dollars[household$Region ==
    1 & household$Area == "Rural"]
```

Numerical summaries can be computed using the tapply() function, but this time instead of giving only one factor to the INDEX= argument, we provide a list of factors using the list() function. The use of list() in INDEX= allows conditioning on multiple factors. Each factor is provided as an additional argument to the list() function.

The table() function can be used to obtain sample sizes, for the combination of region and area. Command Snippet 5.10 provides the syntax to obtain the 20% trimmed means, the 20% Winsorized standard deviations, and the sample sizes for each of the 14 groups.

Notice that the output for the urban area of the Central Coastland region is the value NA. The sample size for this region and area combination is 0, meaning there are no scores for which to compute the summary statistics. This can be seen from the output of the table() function. This can also be seen in Figure 5.7 in which there is no plot for urban households in this region. All the summaries can again be placed in a table for publication as shown in Table 5.2.

Table 5.2: Mean, Standard Deviation and Sample Sizes for Household per Capita Expenditures (in dollars) Conditioned on Urbanicity and Region

Region	Rural			Urban		
	M_t	SD_W	n	M_t	SD_W	n
Central Coast	138	43	502	243	92	252
Central Highlands	126	45	368	—	—	0
Mekong Delta	143	40	830	262	104	282
North Coast	128	35	600	242	98	108
Northern Uplands	109	29	672	215	52	187
Red River Delta	145	37	783	291	107	392
South East	208	66	514	406	150	509

[a] The mean was trimmed by 20%. The standard deviation was computed from a variance that was winsorized by 20%.

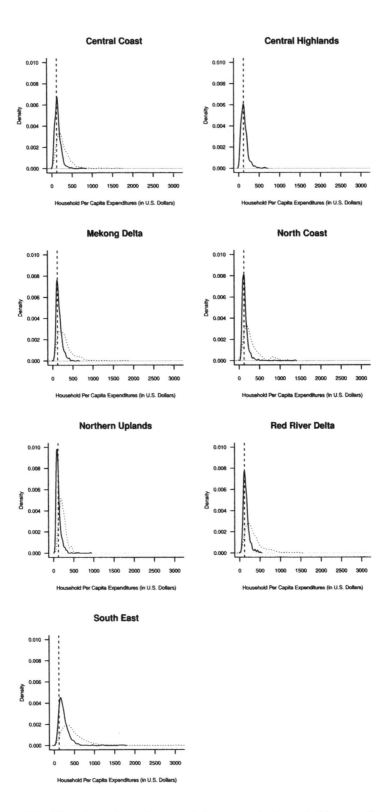

Figure 5.7: Plotted density estimates of the per capita household expenditures for the urban households (solid line) and rural households (dotted line) conditioned on region. The poverty line of $119 is expressed as a vertical dashed line in each panel.

A write-up of the complete results might be as follows.

Sample Write-Up

Given the economic differences that were found in both region and area, a decision was made to examine the differences between rural and urban areas within each region. Table 5.2 shows that in Vietnam, rural households have approximately half the per capita expenditures on average as do urban households. While there is much more variation in the wealth of these urban households, it is apparent that the rural households in every region are not only poorer on average, but also show very little variability.

From examining both the graphical and statistical evidence, there are sample differences in the household per capita expenditures between the regions. The South East region has the highest household per capita expenditure. The distribution of expenditures per capita in the South East region, which includes Ho Chi Minh City and the hinterland, is typically higher and more variable, compared with other regions. Also, more than 75% of the people sampled from the South East region are above the poverty line of $119. The distributions in the other six regions are fairly similar. It is worthwhile to note that in the two most impoverished regions, the Central Highlands and Northern Uplands, close to 50% of the households are below the poverty line.

Command Snippet 5.9: Syntax used to plot the density estimates of the per capita household expenditures for both the urban and rural households within each panel. For the remaining six panels, the object names associated with each region would be used.

```
## Panel 1 - Plot the density for rural households
> plot(density(central.coast.rural), main = "Central Coast",
    xlab = my.xlab,  bty = my.bty, xlim = my.xlim, ylim =
    my.ylim)

## Plot the density for urban households
> lines(density(central.coast.urban), lty = "dotted")

## Add the poverty line
> abline(v = 119, col = "red", lty = "dashed")
```

5.5 EXTENSION: CONDITIONING ON CONTINUOUS VARIABLES

In Chapter 4 and also earlier in this chapter, we examined distributions conditioning on one or more variables. In all of the examples, the conditioning variables were categorical in nature. It is also possible to condition the response variable on factors that are ordinal in nature, or even on continuous variables. To illustrate this type of conditioning, another data set is considered. The data set is from a study that had the primary research question of whether there are differences in mathematics achievement for students who spend differing amounts of time on mathematics homework.

Command Snippet 5.10: Numerical summaries for household per capita expenditures conditioned on region and area.

```
## Conditional 20% trimmed means
## The use of list() in INDEX= conditions on region and area
> tapply(X = household$Dollars, INDEX = list(household$Region,
  household$Area), FUN = mean, tr = 0.2)
    Rural    Urban
1 138.2698 243.1534
2 126.1242      NA
3 142.7892 262.1644
4 127.6951 242.0429
5 108.7761 214.8263
6 144.9908 290.5272
7 208.3707 405.7063

## 20% Winsorized standard deviations conditioned on region and
     area
> sqrt(tapply(X = household$Dollars, INDEX =
  list(household$Region, household$Area), FUN = winvar, tr =
  0.2))
    Rural    Urban
1 42.54099  91.98587
2 44.81304       NA
3 40.04903 103.79783
4 34.53080  97.55267
5 29.49418  51.81709
6 36.83317 107.21399
7 65.60108 150.27420

## Sample sizes conditioned on region and area
> table(household$Region, household$Area)

    Rural Urban
1    502   252
2    368     0
3    830   282
4    600   108
5    672   187
6    783   392
7    514   509
```

During the spring of 1988, the National Center for Education Statistics initiated a longitudinal study—the National Educational Longitudinal Study (NELS)—of eighth-grade students attending 1052 high schools across the 50 states and the

District of Columbia. These students, who constituted a nationally representative sample, were surveyed on a variety of topics related to education, and also given achievement tests in four content areas—reading, social studies, mathematics, and science. Samples of the original 27,394 participants were also surveyed and assessed again in 1990, 1992, 1994, and 2000.[1] The data in *MathAchievement.csv* is a pared down sample from the original 1988 data that consists of two variables— the average amount of time (in hours) spent on mathematics homework per week, and the student's mathematics achievement score for 100 students. The achievement scores are T-scores, which are based on a transformation resulting in a distribution with a mean of 50 and a standard deviation of 10.

The data are read in and the marginal distribution of mathematics achievement is examined. The distributions of mathematics achievement conditioned on the time spent on mathematics homework are also examined. Command Snippet 5.12 shows this syntax. The density plot of the marginal distribution of mathematics achievement, not shown, indicates the estimated density for the mathematics achievement scores is roughly symmetric. As expected for T-scores, the typical value is around 50. There is also variation in the scores.

Side-by-side box-and-whiskers plots are used to examine the distribution of mathematics achievement conditioning on the amount of time spent on homework. When the levels of the factor are not provided in a codebook, the `table()` function returns the levels in addition to the sample sizes corresponding to those levels. Command Snippet 5.12 shows the syntax used to determine the levels of `Homework`. Then indexing is used to create the subsets of mathematics achievement scores at these levels, and box-and-whiskers plots of the conditional distributions are created using the `boxplot()` function. The optional argument `at=` is included in the `boxplot()` function to provide a vector of locations where the box-and-whiskers plots are to be drawn. Without this argument, the box-and-whiskers plots would be drawn at the locations 1 to 8. The limits on the *x*-axis are also changed to include all of the box-and-whiskers plots in the plotting area. Figure 5.8 shows the resulting plot.

The side-by-side box-and-whiskers plots, shown in Figure 5.8, and the numerical summaries suggest that the conditional distributions of mathematics achievement are not extremely asymmetric; and all have roughly the same amount of within-group variation. We also see evidence of differences in mathematics achievement for the differing amounts of time spent on mathematics homework. The side-by-side box-and-whiskers plots show that the typical mathematics achievement score is increasing across the conditional distributions. They also show that the variation in mathematics achievement scores are roughly similar for the differing amounts of time spent on mathematics homework. The small amount of variation shown in the box-and-whiskers plots on the right-hand side of the plot (i.e., at the values of 6, 7, and 10) is likely an artifact of the extremely small sample sizes at those values.

---

[1]For more information on the methodology used in the NELS study, visit the official website: `http://nces.ed.gov/surveys/NELS88/`.

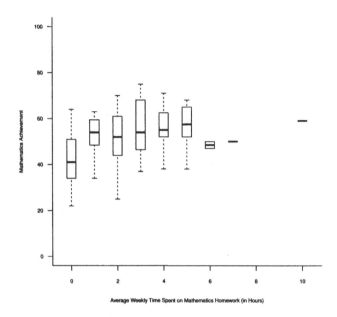

Figure 5.8: The distribution of mathematics achievement conditioned on average weekly time spent on mathematics homework. Achievement is reported as a T-score which are transformed scores having a mean of 50 and a standard deviation of 10.

5.5.1   Scatterplots of the Conditional Distributions

When the conditioning variable is continuous or is measuring a trait that is assumed to be continuous, it can be useful to use a *scatterplot* to examine the conditional distributions. The plot() function is used to produce a scatterplot. To produce a scatterplot, the function requires the arguments x= and y= be specified, with x being the conditioning variable and y being the response variable. Command Snippet 5.11 uses plot() to produce the scatterplot displayed in Figure 5.9.

Command Snippet 5.11: Syntax to plot the conditional distributions of mathematics achievement using a scatterplot.

```
> plot(x = nels$Homework, y = nels$Achievement, xlab = "Average
    Weekly Time Spent on Mathematics Homework (in Hours)",
    ylab = "Mathematics Achievement")
```

Figure 5.9 shows a scatterplot of the relationship between time spent on mathematics homework and mathematics achievement scores for the sample of 100 eighth-

Command Snippet 5.12: Syntax to read in and examine the NELS data, plot the marginal distribution of mathematics achievement scores, determine the levels associated with the Homework variable, use indexing to create the subsets of achievement scores at these levels, and plot the conditional distributions.

```
## Read in the NELS data
> nels <- read.table("/Documents/Data/NELS.csv", header = TRUE,
    sep = ",", row.names = "ID")

## Examine the data frame
> head(nels)
> tail(nels)
> str(nels)
> summary(nels)

## Plot the marginal distribution
> plot(density(nels$Achievement), main = " ", xlab =
    "Mathematics Achievement")

## Determine sample sizes and levels of the factor
> table(nels$Homework)

 0  1  2  3  4  5  6  7 10
19 19 25 16 11  6  2  1  1

## Indexing to create subsets of mathematics achievement
## scores at each level
> zero <- nels$Achievement[nels$Homework == 0]
> one <- nels$Achievement[nels$Homework == 1]
> two <- nels$Achievement[nels$Homework == 2]
> three <- nels$Achievement[nels$Homework == 3]
> four <- nels$Achievement[nels$Homework == 4]
> five <- nels$Achievement[nels$Homework == 5]
> six <- nels$Achievement[nels$Homework == 6]
> seven <- nels$Achievement[nels$Homework == 7]
> ten <- nels$Achievement[nels$Homework == 10]

## Plot the conditional distributions
> boxplot(zero, one, two, three, four, five, six, seven, ten,
    at = c(0:7, 10), xlim = c(-0.4, 10.4))
```

grade students. A point is represented by an open circle and is defined by a person's achievement score and hours of homework. The scatterplot shows the same overall increasing trend in typical mathematics achievement scores as the side-by-side box-and-whiskers plots in Figure 5.8. It also shows the same pattern of variation in the conditional distributions.

Figure 5.10 shows the same scatterplot with side-by-side boxplots superimposed on the plot. This is easily accomplished by plotting the scatterplot and then adding the optional argument add=TRUE to the boxplot() function. This argument adds the side-by-side box-and-whiskers plots to the already existing plot. Command Snippet 5.11 illustrates the syntax that produced Figure 5.10. Superimposing side-by-side boxplots on the scatterplot is a nice way to display smaller data sets. The side-by-side boxplots allow the examination of each conditional distribution, while the scatterplot still shows the individual cases.

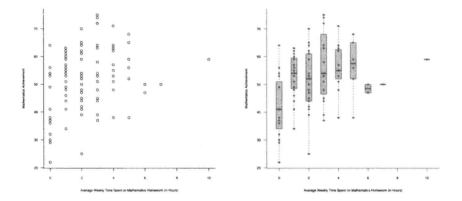

Figure 5.9: Scatterplot showing the distribution of mathematics achievement conditioned on average weekly time spent on mathematics homework.

Figure 5.10: Scatterplot and side-by-side box-and-whiskers plots showing the distribution of mathematics achievement conditioned on average weekly time spent on mathematics homework.

Command Snippet 5.13: Syntax to plot the conditional distributions of mathematics achievement using a scatterplot.

```
## Plot the data using a scatterplot
## The limits on the x-axis are changed
## pch= changes the point character to solid dots
> plot(x = nels$Homework, y = nels$Achievement, xlab = "Average
    Weekly Time Spent on Mathematics Homework (in Hours)",
    ylab = "Mathematics Achievement", xlim = c(-0.4, 10.4), pch
    = 20)

## Add the side-by-side box-and-whiskers plots
## boxwex= changes the width of the boxes
## Transparent color is added to enhance the plot
> boxplot(zero, one, two, three, four, five, six, seven, ten,
    at = c(0:7, 10), add = TRUE, axes = FALSE, boxwex = 0.4,
    col = rgb(red = 0.2, green = 0.2, blue = 0.2, alpha = 0.3))
```

5.6 FURTHER READING

Panel plots were introduced by Cleveland (1993) as a part of a data visualization framework called *trellis graphics* that was developed at Bell Labs. A more extensive implementation of the trellis graphics framework in R is the **lattice** package. Sarkar (2008) provides an excellent starting point for the reader interested in learning more about this package. For more information on the use and interpretations of scatterplots, any good regression book would suffice (e.g., Cook & Weisberg, 1999; Mosteller & Tukey, 1977; Sheather, 2009).

PROBLEMS

5.1   Using the *LatinoEd.csv* data set, plot the distribution of `Achieve` conditioned on `ImmAge` in a scatterplot. Also, include side-by-side box-and-whiskers plots. Based on the plot, describe the relationship between age at immigration and level of educational achievement for these Latino immigrants. Be sure to examine the pattern in variation as well as the pattern in locations. Write up the results of your investigation as if you were writing a manuscript for publication in a journal in your substantive area.

5.2   There are two variables in the *LatinoEd.csv* data set which are dichotomously scored—`English` and `Mex`. Together these variable represent four distinct subgroups of the sampled Latino immigrants: (1) fluent in English—immigrated from Mexico; (2) fluent in English—didn't immigrate from Mexico; etc. Provide both graphical and numerical summaries to compare the educational achievement of these four groups. Write up the results as if you were writing a manuscript for publication in a journal in your substantive area. Be sure to comment on any interesting facets of the data when they are examined in this manner.

CHAPTER 6

RANDOMIZATION AND PERMUTATION TESTS

... it is only the manipulation of uncertainty that interests us. We are not concerned with the matter that is uncertain. Thus we do not study the mechanism of rain; only whether it will rain.

—D. Lindley (2000)

In Chapter 4, differences between two groups were examined. Specifically, the question of whether there were differences in the annual household per capita expenditures between the rural and urban populations in Vietnam was addressed. In that chapter, exploratory methods, such as graphical and numerical summarizations, were used to quantify the differences in the two distributions of household per capita expenditures. Exploration is often only the starting point for examining research questions involving group differences. These methods, however, do not always provide a complete answer to the research question. For example, most educational and behavioral researchers also want to determine whether the differences that might have shown up in the exploration phase are "real," and to what population(s) the "real" effect can be attributed. A "real" effect is a sample effect that is caused by an actual difference in the population of interest. For example, suppose the mean per capita household expenditures for the entirety of Vietnam is actually less for rural regions. Then a

Comparing Groups: Randomization and Bootstrap Methods Using R
First Edition. By Andrew S. Zieffler, Jeffrey R. Harring, & Jeffrey D. Long

sample result would be expected to reflect this, provided the sample was obtained in a particular way, namely, randomly (see below). In addition to evaluating whether effects are "real", it is important to estimate the size of the effect. Uncertainty is always involved in this endeavor, which relates to the *precision* of the estimate.

Questions of whether or not group differences are "real", estimates of the size of group differences, and the precision of these estimates are typically problems of *statistical inference*. In the next several chapters, some useful methods to answer these types of inferential questions are introduced. First, however, two research questions regarding group differences that have been studied by educational and behavioral scientists are presented.

Research Question 1 Demands for accountability and delinquency prevention in recent years have led to rising popularity of after-school programs in the United States. The intuitive appeal of these programs is based on the perception that adolescents left unsupervised will either simply waste time or, worse, engage in delinquent and dangerous behaviors.

To empirically study the effects of attending an after-school program, Gottfredson, Cross, Wilson, Rorie, and Connell (2010) randomly assigned middle-school students to either a treatment group or control group. The treatment consisted of participation in an after-school program, whereas the control group engaged in their usual routine, but control students were invited to attend one after-school activity per month. Data on several outcome measures were collected on the study participants. These data are available in *AfterSchool.csv*. The researchers were interested in determining whether there is a difference in the effect of delinquency between students who participated in the after-school program and students that did not.

Research Question 2 The Center for Immigration Studies at the United States Census Bureau has reported that despite shifts in the ethnic makeup of the immigrant population, Latin America—and Mexico specifically—remains this country's greatest source of immigrants. Although the average immigrant is approximately 40 years old, large numbers are children who enroll in U.S. schools upon arrival. Their subsequent educational achievement affects not only their own economic prospects but also those of their families, communities, and the nation as a whole.

Stamps and Bohon (2006) studied the educational achievement of Latino immigrants by examining a random sample of the 2000 decennial Census data, a subset of which is provided in LatinoEd.csv. One interesting research question that has emerged from their research is whether there is a link between where the immigrants originated and their subsequent educational achievement. Specifically, the question is if there is a difference in the educational achievement of immigrants from Mexico and that of immigrants from other Latin American countries.

Random Assignment & Random Sampling While both of these research questions may seem similar—apart from their context—they are in fact very different. In the first situation, the researchers used a volunteer sample and randomly assigned the participants in their sample to the two groups—treatment and control. In the second situation, the researchers randomly selected their sample from a larger population (the 2000 census), but the two groups were not assigned by the researchers. These examples illustrate two important differences: (1) how the sample is selected and

(2) how the treatments, or groups, are assigned. Table 6.1 shows the four potential situations that educational and behavioral science researchers could face.

Table 6.1: Four Potential Scenarios Researcher Could Face When Making Inferences

Scenario	RS[a]	RA[b]	Type of Research
Scenario 1	✓		Generalizable research
Scenario 2		✓	Randomized experimental research
Scenario 3	✓	✓	Generalizable, randomized experimental research
Scenario 4			Nongeneralizable, nonexperimental research

[a] RS = Random sample

[b] RA = Random assignment

Each of these scenarios impacts the conclusions that a researcher can draw from quantitative results. How the sample is selected has a direct impact on the generalizations that a researcher can draw. For example, random sampling helps ensure that the conclusions drawn from the sample data can be generalized to the population from which the sample was drawn. In contrast, how the treatments are assigned has a direct impact on the causal inferences a researcher can make. Random assignment to treatments facilitate these causal inferences by allowing the attribution of sample differences to the differences in treatments.

The nomenclature used in Table 6.1 is employed in this monograph to help educational and behavioral researchers make clearer distinctions between these scenarios. Unfortunately, the discipline of statistics uses no consistent terminology to describe each of these scenarios. For example, in research that employ random assignment, the term "experiment" is sometimes used, but as Kempthorne (1979, p. 124) points out:

> The literature of statistics has been plagued with difficulties about the use of the word "experiment." Interestingly enough, so also has the general world of science. The problem is that a loose use of the word "experiment" permits it to be applied to any process of examination of a space-time section of the world.

The use of random sampling and/or random assignment are the components that allow statistical inference to take place. The mathematical theory for inferential methods is, in fact, intrinsically tied to the employment of one or both of these random mechanisms. In this chapter, methods that allow researchers to answer research questions in which the researcher has randomly assigned the treatments are examined. In Chapter 7, methods that allow researchers to answer research questions if the study used random sampling to select the sample are examined. The other two scenarios—generalizable, randomized experimental research and nongeneralizable research—are touched on in both Chapters 6 and 7 and are discussed further in Chapter 8.

6.1 RANDOMIZED EXPERIMENTAL RESEARCH

Consider researchers who are studying whether after-school programs have an effect on delinquency. What if they observed that students who participate in after-school programs tended to have a low measure of delinquency? What could be said about the effect of after-school programs on delinquency? Should the researchers conclude that after-school programs lessen delinquency for students? That, of course, is one explanation for the observed relationship. However, there are several alternative explanations as well. One rival explanation is that students with a low propensity for delinquency to begin with were the ones who participated in the after-school program. And, had these students not participated in the after-school program, they still would have had low measures of delinquency. Because there is no comparison group—no variation in the treatment predictor—there is no way to establish which explanation is correct.

When examining the effect of a treatment or intervention, it is essential that educational and behavioral researchers specify a comparison group—often referred to as a *control group*. The comparison group attempts to answer the question, what would have happened to the group of students if they did not receive the treatment? Without a comparison group, it is impossible to rule out alternative explanations for the "effect" that is being examined. In fact, because the conclusions drawn from research that is conducted without comparison groups are relatively weak, many experts suggest that such research is a waste of time and "at best, offer indecisive answers to the research questions, and, at worst, might lead to erroneous conclusions" (Light, Singer, & Willett, 1990, pp. 104–105)."

What comparison group should the after-school program researchers choose? Perhaps they should compare the students who participated in the after-school program to other students who did not participate in the program. Should the students in the comparison group be from the same school as those in the treatment group? Or maybe from the same neighborhoods? Another option is to use the same students in the comparison and treatment group. The researchers could compare these students' delinquency measures both before and after they participated in the after-school program. This type of design is often referred to as a *pre–post design* and will be examined in more detail in Chapter 10. Each of these comparison groups would yield a different answer to whether or not the after-school program has an effect on delinquency. Furthermore, depending on how the comparison group is selected, there may still be alternative explanations that cannot be ruled out, because the apparent effects could be due to attributes or characteristics, called *confounding variables*, that are systematically related to the treatment (e.g., socioeconomic status, scholastic engagement, etc.).

Because of the potential problems with confounding variables, the choice of a comparison group is an important task in research design. The best comparison group is one that is "composed of people who are similar to the people in the treatment group in all ways except that they did not receive the treatment" (Light et al., 1990, p. 106). The use of random assignment, or randomization, to assign students to *conditions* is a statistical means of simultaneously considering all of the potential confounding variables, both known and unknown. The word *condition* is a generic term that includes the comparison group and the treatment group. Specifically, random assigment of participants to conditions (or conditions to participants) is a method that ensures that the treatment group and the comparison group are equivalent, *on average*, for all attributes, characteristics, and variables other than the treatment.

Again, consider the after-school program research study described in the previous section. Because the researchers randomly assigned middle-school students to either a treatment group or control group, the two groups of students should be equivalent, on average, for any variable that might be related to the treatment. Because of the randomization, any effects or differences in delinquency that the researchers find, must be due to the variation in conditions because the equivalence induced by the random assignment rules out other explanations. Variation in conditions in this context means the students either participated in an after-school program or they did not.

6.2 INTRODUCTION TO THE RANDOMIZATION TEST

Randomization tests are statistical tests that can be used to evaluate hypotheses about treatment effects when experimental units have been randomly assigned to treatment conditions. To help illustrate the concept of the randomization tests, a pedagogical example is considered. Imagine a counseling psychologist was interested in determining if there was an effect of cognitive-behavioral and social problem-solving training on perceived social competence in school-aged aggressive boys.[1] The researcher randomly assigned three aggressive boys to a control group and two others to a treatment group who received both cognitive-behavioral and social problem-solving training. After the study, a scale of perceived social competence was administered to all five participants, and their scores, respectively, were

Treatment (T): 54, 66   Control (C): 57, 72, 30,

with higher scores indicating a higher degree of perceived social competence. Does the higher average measure of perceived social competence in the treatment group— 60 versus 53—provide convincing evidence that the training is effective? Is it possible that there is no effect of training, and that the difference observed could have arisen just from the nature of randomly assigning the five participants into groups? After all, it cannot be expected that randomization will always create perfectly equal groups. But, is it reasonable to believe the random assignment alone could have lead to this large of a difference? Or is the training also contributing to this difference?

To examine the above questions, one approach is to imagine the scenario under which the training had no effect whatsoever. One can go even further and always consider this scenario to be the default scenario. The purpose of statistical inference is to evaluate the default scenario. If there is sufficient evidence against the default

---

[1]This example was inspired by a similar study carried out by Lochman, Lampron, Gemmer, Harris, and Wyckoff (1989).

scenario, then it would be discarded as implausible. If there is insufficient evidence against the default scenario, then it would not be discarded. Specifically, in this example, an assumption would be made that the training has no effect. Then, evidence would be collected to determine if the difference that was observed in the data is too large to probabilistically believe that there is no effect of training. This statement or assumption of no treatment effect is called the *null hypothesis* and is written as

$$H_0 : \text{The training is not effective.}$$

If the training is truly ineffective, then each participant's perceived social competence score is only a function of that person and not a function of anything systematic, such as the training. The implication of the participant's perceived social competence score not being a function of anything systematic is that, had a participant been assigned to the other condition (through a different random assignment), his perceived social competence score would have been identical since, in a sense, both conditions are doing nothing in terms of affecting the perceived social competence scores.

One can take advantage of the fact that each participant's perceived social competence score would be identical whether they are assigned to treatment or control and examine *all possible* random assignments of the participants to conditions. Table 6.2 shows all 10 possible permutations (i.e., arrangements) of the data, as well as the mean difference in perceived social competence scores for those assignments. The notation $\bar{T}$ and $\bar{C}$ are used for the mean of the treatment group and mean of the control group, respectively. The term *permutation* here refers to a unique rearrangement of the data that rises from random assignment.[2] Mathematically, there are

$$\binom{5}{2} = \frac{5!}{2!(5-2)!} = 10,$$

such rearrangements, where $5! = (5)(4)(3)(2)(1)$, and similarly for the other values. In general, the number of unique permutations of n measurements into samples of size k and $n - k$ is computed using

$$\binom{n}{k} = \frac{n!}{k!(n-k)!}, \tag{6.1}$$

where $n! = (n)(n-1)(n-2)\ldots(1)$.

The result that was observed, a difference of 7 points, is of course one of the possibilities under the assumption that there is no effect of training. The big question is whether or not it is likely that an observed mean difference of 7 points is large enough to say that it is due to something (i.e., the training) affecting the perceived social competence scores, or is it simply an artifact of the random assignment? Typically, educational and behavioral researchers provide a quantification of the strength of evidence against the null hypothesis called the *p*-value, which helps them

---

[2]Note that the term permutation as used in this context is different from the strictly mathematical definition, which is a reordering of the numbers $1, \ldots, n$ and is computed as $n!$ (n factorial).

Table 6.2: Ten Possible Permutations of the Perceived Social Competence Scores and Difference in Mean Perceived Social Competence Scores

Score	Treatment	Control	$\bar{T} - \bar{C}$
1	54, 66	57, 72, 30	$60 - 53 = 7$
2	54, 57	66, 72, 30	$55.5 - 56 = -0.5$
3	54, 72	66, 57, 30	$63 - 51 = 12$
4	54, 30	66, 72, 57	$42 - 65 = -23$
5	57, 66	54, 72, 30	$61.5 - 52 = 9.5$
6	72, 66	57, 54, 30	$69 - 47 = 22$
7	30, 66	57, 72, 54	$48 - 61 = -13$
8	57, 72	54, 66, 30	$64.5 - 50 = 14.5$
9	57, 30	66, 72, 54	$43.5 - 64 = -20.5$
10	72, 30	66, 54, 57	$51 - 59 = -8$

answer this question. To compute the p-value, the proportion of random permutations of the data that provide a result *as extreme or more extreme* than the one observed is computed. Nine of the 10 potential results are as extreme or more extreme than a 7-point difference. Mathematically, this is written as

$$P\left(\left|\text{observed difference}\right| \geq 7\right) = \frac{9}{10},$$

where $|\cdot|$ is the absolute value.

If there is no effect of training, 9 out of the 10 permutations of the data that are possible would produce a result as extreme or more extreme than a 7-point difference. The p-value obtained from a study is a piece of evidence that can be used to evaluate the tenability of the null hypothesis. Smaller p-values provide stronger evidence against the null hypothesis. For example, the p-value of 0.5 provides very weak evidence against the null hypothesis that training is ineffective.

Many educational and behavioral researchers use a p-value to make a decision about the null hypothesis. This is a very bad idea. For one thing, p-values are tremendously impacted by the size of the sample. For example, in this example, the psychologist who was evaluating whether or not the training was effective had conditions in which the sample sizes were $n_1 = 2$ and $n_2 = 3$. The p-value could have been large simply because the sample sizes were so small. With such little data, the number of possible permutations is limited, meaning the amount of information for evaluating the null hypothesis is also limited.

Secondly, decisions about null hypotheses often impact lines of research, and it is unclear if the results from one study should carry such weight. The null hypothesis test was developed by R. A. Fisher, who initially proposed the p-value as an informal index to be used as a measure of discrepancy between the observed data and the null hypothesis being tested, rather than a part of a formal inferential decision-making process (Fisher, 1925). He went on to suggest that p-values be used as part of the fluid, nonquantifiable process of drawing conclusions from observations, a process that included combining the p-value in no one specific manner with field substantive information and other research evidence. Above all, one can say that Fisher's emphasis was on *replication* of results. Only if a result was replicated could it possible carry sufficient weight to impact a line of research.

What was considered above is one example of a randomization test. As the plural, randomization tests, in the chapter title implies, this name refers to a class of procedures, and not a single test or procedure. All randomization tests, however, are derived from the same fundamental principle. Namely, that the reference distribution of a particular statistic can be obtained by assuming no differences between the groups. Under this assumption, the statistic of interest, such as the mean difference, is calculated under all the potential random assignments of the treatment conditions on the observed data. Then the reference distribution is used to evaluate the likelihood of seeing a result as extreme or more extreme than the one observed in the original data.

6.3 RANDOMIZATION TESTS WITH LARGE SAMPLES: MONTE CARLO SIMULATION

In the last section, the p-value calculated from the randomization test was *exact*. It was exact because *all possible* permutations of the data were used in the calculation of its value. When $N = 5$, it is possible to list all of the permutations of the data. However, for larger sample sizes, the computation of all the permutations becomes an impractical or even impossible task. To illustrate, the data from the after-school research study introduced in Research Question 1 (see above) is examined to determine whether there is a difference in the effect of delinquency between students who participated in an after-school program and students that did not. These data can be found in *AfterSchool.csv*.

After an initial inspection of the data codebook, the data can be read into R and both graphical and numerical summaries can be examined. Command Snippet 6.1 shows the syntax. Henceforth, only selected exploratory results from the R Command Snippets will be presented.

Figure 6.1 shows the kernel density plots for the distribution of the standardized delinquency measure conditioned on treatment condition.

Examination of Figure 6.1 reveals similarities in the distributions of the two groups. The numeric summaries indicate there are slight differences between the treatment and control groups. The sample means suggest that the difference in T-scaled delinquency measures between these groups is, on average, 1.5 ($50.7 - 49.0 = 1.7$).

Is this difference within the chance variation one would expect given random assignment? Or is it outside of those expectations? In other words, what is the

Command Snippet 6.1: Syntax to read in and examine the after-school data.

```
## Read in the data
> asp <- read.table(file = "/Documents/Data/AfterSchool.csv",
    header = TRUE, sep = ",", row.names = "ID")

## Examine the data frame object
## Output is suppressed
> head(asp)
> tail(asp)
> str(asp)
> summary(asp)

##Exploratory density plot for treatment group
> plot(density(asp$Delinq[asp$Treatment == 1], bw = 3), main=
    " ", xlab= "T-Scaled Delinquency Measure", bty = "l", col =
    "#377EB8", lty = "solid")

##Exploratory density plot for control group
> lines(density(asp$Delinq[asp$Treatment == 0], bw = 3), col =
    "#E41A1C", lty ="dashed")

## Add legend
> legend(x = 70, y = 0.085, legend = c("Control Group",
    "Treatment Group"), col = c("#E41A1C", "#377EB8"), lty =
    c("dashed", "solid"), bty = "n")

## Conditional means
> tapply(X = asp$Delinq, INDEX = asp$Treatment, FUN = mean)
         0         1
50.72559 49.01896

## Conditional standard deviations
> tapply(X = asp$Delinq, INDEX = asp$Treatment, FUN = sd)
         0         1
10.52089  8.97423

## Sample sizes
> table(asp$Treatment)

  0   1
187 169
```

expected mean difference if there is no effect of after-school programs and different students had been randomly assigned to the treatment and control groups? For this example, there are over 3.93×10^{105} permutations of the data! This would take a very long time indeed to list out the possible rearrangements of the data. For this reason, researchers use *Monte Carlo simulation* to approximate the *p*-value in situations where the number of permutations is prohibitively time consuming to list.

Monte Carlo simulation is a method that uses a much smaller random sample (say 5000) of all of the random permutations to approximate the reference distribution of the test statistic under inquiry. The approximation is generally very good, and thus, approximate methods are in wide use. At the heart of Monte Carlo methods is the simple idea of selecting a statistical sample to approximate the results rather than to work out the often much more complicated exhaustive solution.[3] Dwass (1957) used the Monte Carlo method to simplify the problem of examining all permutation results and found that it provided a close match to the exact results.

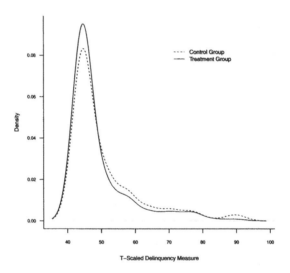

Figure 6.1: Kernel density plots for the distribution of the T-scaled delinquency measure for students who participated in the after-school program (solid line) and for students who did not (dashed line).

6.3.1 Rerandomization of the Data

A random permutation of data can be drawn using the `sample()` function. This function samples the original scores *without replacement*. It is comparable to writing each of the delinquency scores on a notecard, shuffling those cards, and then dealing them out into a new order. This results in a random rearrangement of the original delinquency scores. Command Snippet 6.2 shows the syntax for assigning the randomly permuted delinquency measures to an object called `permuted`.

How does one know which of the scores in `permuted` will be "assigned" to the control group and which to the treatment group? Recall from the exploratory analysis

---

[3]In fact, the method was initially proposed by Stanislaw Ulam, in 1946, who wanted to know the probability that a Canfield solitaire laid out with 52 cards would be sucsessful (Eckhardt, 1987). After trying in vain to solve the problem exhaustively through mathematical combinatorics, Ulam laid out several random deals and counted the number of successes.

that there are 187 participants in the original control group and 169 participants in the treatment group. Replicating the same sizes with our permuted data is the most important part of the problem—so long as this assignment is consistent from permutation to permutation. Thus, the first 187 scores in the vector can be assigned to the control group and the remaining 169 scores to the treatment group. Command Snippet 6.3 shows the use of indexing to compute the mean difference for the permuted control scores and the permuted treatment scores.

Command Snippet 6.2: Syntax to randomly permute the delinquency measures and assign them to an object called `permuted`.

```
## Randomly permute the delinquency measures
> permuted <- sample(asp$Delinq)

## Examine the vector of permuted measures
> head(permuted)
[1] 44.46308 76.81361 44.46308 57.40329 76.81361 44.46308

> tail(permuted)
[1] 44.46308 44.46308 44.46308 57.40329 44.46308 50.93319

> summary(permuted)
   Min. 1st Qu.  Median    Mean 3rd Qu.    Max.
  44.46   44.46   44.46   49.92   50.93   89.75
```

Command Snippet 6.3: Syntax to find the mean difference for the permuted control scores and the permuted treatment scores.

```
> mean(permuted[1:187])  - mean(permuted[188:356])
[1] -0.4070047
```

6.3.2 Repeating the Randomization Process

To obtain a Monte Carlo *p*-value, many random permutations of the data will need to be drawn. For each permutation the mean difference will also need to be computed. Statistical computing tasks often require repeated computations. This can be performed in a number of ways in R. Use of the `replicate()` function is one method in which to automate such tasks. This function applies a computation repeatedly a given number of times and collects the results into a vector or matrix, whichever is convenient. There are two arguments used with the function, the number of replications to perform using n=, and the computation to carry out in the argument expr=. Command Snippet 6.4 shows the use of the `replicate()` function to carry out 4999 random permutations of the delinquency scores.

The assigned object `permuted` is a 356×4999 matrix, where each column contains one of the 4999 random permutations of the data. It is now desirable to compute the difference between the mean of the first 187 scores and the mean of the remaining 169 scores for each column. To do this, an R function is written to compute this mean difference. This new function is then executed on all 4999 columns of the matrix. The next section describes how to write simple functions in R.

Command Snippet 6.4: Syntax to carry out 4999 random permutations of the delinquency scores and store them in the object permuted.

```
> permuted <- replicate(n = 4999, expr = sample(asp$Delinq))
```

6.3.3   Generalizing Processes: Functions

Functions are the fundamental structure of "modern" programming. A function is a way of organizing operations and computations to allow them to be used over and over again without duplicating the commands themselves. To define a new function in R the function operator will be utilized. The basic syntax for this operator follows the following pattern.

```
> function(argument, argument, ...){
    expression 1;
    expression 2;
        ⋮
  }
```

Functions are typically assigned to an object, which then becomes the name of the function. The *arguments* are the inputs that the function will take, each being separated by a comma. The *expressions*—which are enclosed in braces[4]—make up the body of the function. Each expression, or computation, needs to be separated by a semicolon, or a carriage return. In Command Snippet 6.5 a function, called mean.diff(), is written to compute the mean difference between the first 187 scores and the remaining 169 scores for a vector of length 356.

Command Snippet 6.5: A function to compute the mean difference between the first 187 scores and the remaining 169 scores for a vector of length 356.

```
mean.diff <- function(data) {
  mean(data[1:187]) - mean(data[188:356])
  }
```

---

[4]If the function only contains a single expression, it does not have to be enclosed in braces, but it is good practice to do so anyway.

By not specifying the argument data as a particular value, the function will be able to use any vector of scores that is supplied as an input. After executing the syntax from Command Snippet 6.5 in R, the function can be used just like any other function. In Command Snippet 6.6 the new function, mean.diff(), is used to find the mean difference between the first 187 and last 169 permuted delinquency scores in the observed data. Since the data are ordered such that the first 187 scores belong to the control group and the last 169 scores belong to the treatment group, the function returns the observed mean difference. When the data are sorted by the treatment factor, this can be a good way to test that the function is working correctly.

Command Snippet 6.6: Using the mean.diff() function to find the mean difference between the first 187 and last 169 delinquency scores in the observed data.

```
> mean.diff(asp$Delinq)
[1] 1.706636
```

6.3.4 Repeated Operations on Matrix Rows or Columns

The goal is to execute the mean.diff() function on all 4999 columns of the permuted matrix. The apply() function is used to carry out a particular function or computation on each row or column of a matrix or data frame. This function takes the arguments X=, MARGIN=, and FUN=. The first argument indicates the name of the matrix or data frame to be used. The second argument indicates whether the function should be applied to the rows (MARGIN=1), columns (MARGIN=2), or both (MARGIN=c(1, 2)). The last argument, FUN=, is the name of the function that should be applied, in this case FUN=mean.diff. Since the mean.diff() function needs to be carried out on each column in permuted, the arguments X=permuted and MARGIN=2 are used. Command Snippet 6.7 shows the syntax for carrying out this computation and assigning the computed mean differences to a vector object called diffs. Note that mean.diff() is used without the parentheses in the syntax.

Command Snippet 6.7: Syntax to compute the mean difference between the first 169 and last 187 permuted delinquency scores in each column of the object permuted.

```
> diffs <- apply(X = permuted, MARGIN = 2, FUN = mean.diff)
```

6.3.5 Examining the Monte Carlo Distribution and Obtaining the *p*-Value

To examine the mean differences, exploratory methods are utilized on the new vector object, diffs. It is always useful to plot the Monte Carlo distribution to assess if the permutations were correctly carried out. The syntax to plot the density of the permuted mean differences is provided in Command Snippet 6.8, along with the syntax to summarize this distribution.

Command Snippet 6.8: Syntax to compute the mean difference between the first 169 and last 187 permuted delinquency scores in each column of the object permuted.

```
## Plot the density of the permuted mean differences
> plot(density(diffs))

## Numerically summarize the distribution of the permuted mean
     differences
> summary(diffs)
      Min.   1st Qu.    Median      Mean   3rd Qu.      Max.
 -3.468000 -0.698500  0.030300  0.009232  0.759100  3.529000

## Examine the variation of the permuted mean differences
> sd(diffs)
[1] 1.053810
```

The plot of the kernel density estimate—shown in Figure 6.2—indicates that the distribution of the 4999 permuted mean differences is roughly normally distributed. As seen in Figure 6.2, the mean differences are, on average, close to 0, as this value is at the base of the curve peak. This is confirmed by examining the printed summary output. Since the delinquency scores were permuted under the assumption of no difference between the two groups, an average mean difference near zero is expected. The standard deviation of 0.10 is relatively small. This suggests that there is little variation in the mean differences based on only differences in random assignment. Because the standard deviation is quantifying the variation in a statistic (e.g., the mean difference), it is referred to as a *standard error*.

To obtain the Monte Carlo p-value, recall that the proportion of permuted sample mean differences as extreme or more extreme than the original observed difference of 1.7 needs to be computed. The sort() function can be used to sort a vector from the smallest to the largest element. The elements as extreme or more extreme than 1.7 could then be manually counted. This is, of course, quite time consuming—especially when the number of permutations is large.

A better method is to use indexing to identify these elements and the length() function to count them. Command Snippet 6.8 shows the syntax for using indexing to count the elements of diffs that are greater than or equal to 1.7 and also the number of elements that are less than or equal to -1.7. The abs() function can also be used to reduce this to one line by taking the absolute value of the mean differences. This is shown in the third line of Command Snippet 6.9.

Based on the random permutations of data, 577 of the 4999 random permutations resulted in a mean difference as extreme or more extreme than the observed mean difference of 1.7. This would seem to suggest that the p-value is

$$P\left(\left|\text{observed difference}\right| \geq 0.17\right) = \frac{577}{4999} = 0.115.$$

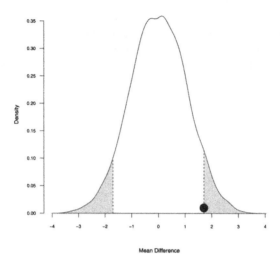

Figure 6.2: Kernel density plot for the distribution of permuted mean differences. The point represents the observed mean difference of 1.7. The shaded area represents the p-value.

This proportion is an estimate of the shaded area in Figure 6.2. It turns out that this method of calculating a p-value gives a biased estimate in that it tends to underestimate the true p-value when using Monte Carlo simulation. This happens because only a random sample of all possible permutations has been taken. Davison and Hinkley (1997) provide a correction to the Monte Carlo p-value as

$$p = \frac{r+1}{k+1},$$ (6.2)

where r is the number of permutations (replicates) that are as extreme or more extreme than the observed statistic, and k is the number of randomly sampled permutations performed in the Monte Carlo simulation. Using the correction,[5] the Monte Carlo p-value for our example would be

$$p = \frac{577+1}{4999+1} = 0.116.$$ (6.3)

Based on either the corrected or uncorrected p-value, the null hypothesis is not discarded. That is, the statistical evidence is not particularly strong to rule out no difference between the treatment and control groups. It is important to note that when

[5]This adjustment is why some researchers prefer to have the sampled number of permutations end in a 9 (e.g., 4999 or 9999).

the number of permutations that are sampled is large, the difference between the value of the corrected and uncorrected *p*-value is virtually nonexistent, and furthermore, that this correction is not universally carried out (Boos, 2003). Aside from producing a less biased estimate of the true *p*-value, this correction also avoids the problem of obtaining a *p*-value of 0 when the observed test statistic is greater than any that were randomly sampled in the simulation, since the minimum possible estimate would be $1/(k+1)$. Because of this, it is a good habit to always use the adjustment in practice.

Command Snippet 6.9: Using indexing to count the elements in my.vector that are greater than or equal to 0.17 and the number of sample differences that are less than or equal to -0.17.

```
## Count the number of permuted mean differences lower than
    -1.7
> length(diffs[diffs <= -1.7])
[1] 270

## Count the number of permuted mean differences higher than
    1.7
> length(diffs[diffs >= 1.7])
[1] 307

## Count the number of permuted mean differences more extreme
    than 1.7
> length(diffs[abs(diffs) >= 1.7])
[1] 577
```

6.4 VALIDITY OF THE INFERENCES AND CONCLUSIONS DRAWN FROM A RANDOMIZATION TEST

The validity of the inferences and conclusions one can draw from any statistical test always need to be evaluated. For example, in order to validly use the *p*-value obtained from a statistical test as a measure of the strength of evidence against the null hypothesis, there are certain criteria or *assumptions* that need to be met. These assumptions vary depending on the statistical test that is performed. For the conclusions from the randomization test to be valid, the assumption of *exchangeability* needs to hold.

6.4.1 Exchangeability

Exchangeability for the randomization test specifies that, under the null hypothesis, all possible permutations of the data are *equally likely*. In other words, each permutation of the data is exchangeable with any other permutation of the data. This assumption is required since the computation of the *p*-value used in the randomization test gives equal weight to each permutation of the data.

In randomized experimental research, whether generalizable or not, the assumption of exchangeability is tenable. The randomization test permutes data in a manner that is consistent with the procedure of random assignment. Thus, the results obtained are valid when random assignment has been initially used in the study. In the after-school example previously considered, exchangeability was assumed to hold, as the participants were randomly assigned to the conditions.

6.4.2 Nonexperimental Research: Permutation Tests

In nonexperimental research studies that do not employ random sampling, the assumption that each permutation of the data is equally likely is a very hard case to make. Some permutations of the data are probably more likely than others. In the absence of random assignment, is it possible to validly draw conclusions from the results of applying this procedure? When the same procedure is applied to data from groups which were randomly sampled rather than randomly assigned, the procedure is called a *permutation test* rather than a randomization test.[6] Oscar Kempthorne, who has written extensively about the theory of randomization, has distinguished between randomization and permutation tests, writing, "the distinction between randomization tests and permutation tests is important. The latter are based on the assumption of random sampling, an assumption that is often patently false or unverifiable, even though necessary to make an attack on the substantive problem being addressed" (Kempthorne, 1986, p. 524).

The use of permutation tests in nonexperimental research is becoming more widespread. When random sampling has been employed, but random assignment to conditions has not been implemented, Pitman (1937) has provided theoretical justification for the use of permutation methods in producing valid results. An example of random sampling without random assignment is the example comparing the educational achievement between Mexican and non-Mexican immigrants.

The rationale for permutation tests, however, is quite different from that for randomization tests. For nonexperimental research with random sampling, the permutations of data do not provide potential outcomes for the same participants, but rather outcomes for other participants that may have been drawn from hypothetical, infinitely sized, populations. In such cases, although the use of the permutation method is justified, the results and conclusions need to be interpreted with caution. It is no longer the case that the results can be used to claim cause and effect arguments.

6.4.3 Nonexperimental, Nongeneralizable Research

More likely, in education, is the situation in which there is no random sampling and no random assignment. In nonexperimental, nongeneralizable research showing that the assumption of exchangeability has been met is a very difficult exercise. It is not supported via random assignment nor through random sampling. Should one not bother analyzing data in which neither random sampling nor random assignment have

---

[6]It is important to note that some authors use the two terms synonymously.

taken place? The authors of this monograph take the position that some information is better than no information, even if the information is less precise or need to be qualified in some fashion. Winch and Campbell (1969) argued that although the results from nonexperimental research in which random sampling did not occur cannot be used to identify causes of observed differences, small *p*-values can help researchers rule out chance differences due to random fluctuations in the data. Such conclusions can be very powerful. They suggest that such conclusions might help a researcher decide whether or not follow-up research might be worth pursuing, or if one should abandon pursuit of the investigation in question. Perhaps the best advice was proffered by Charles Babbage (in Blackett, 1968, p. xiii), who said, "Errors using inadequate data are much less than those using no data at all."

6.5 GENERALIZATION FROM THE RANDOMIZATION RESULTS

In the after-school program example, the data used to examine the effects was obtained from a research design that used random assignment. Based on the design and data collection methods, what are the conclusions that can be drawn and the inferences that can be made regarding after-school programs?

As Campbell and Stanley (1963, p. 5) point out, typically, educational and behavioral researchers are interested in two types of validity questions, namely (1) whether "the experimental treatments make a difference in this specific experimental instance"; and (2) "to what populations, settings, treatment variables, and measurement variables can this effect be generalized." These two questions refer to types of validity—*internal* and *external*—respectively.

The random assignment of students to conditions allows the researchers to draw conclusions about the effects of the treatment in this particular study—it provides evidence of internal validity. It is unlikely that there are average differences between students who participated in after-school programs and those that did not—there is no effect of after-school program on delinquency—for students who participated in the study.

Do the results from this specific study, no treatment effect of the after-school program, hold for different students in different schools in different communities? The answer to this question is unclear. As Edgington and Onghena (2007, p. 8) point out, "statistical inferences about populations cannot be made without random samples from those populations . . . in the absence of random sampling, statistical inferences about treatment effects must be restricted to the participants (or other experimental units) used in an experiment."

Does this imply that the conclusions regarding the effects of treatment can only be made about the 356 students used in this study? Statistically, the answer to this question is "yes." However, *nonstatistical inferences* can be drawn on the basis of logical considerations. Drawing nonstatistical generalizations is a common practice in educational and behavioral research where samples are often not randomly sampled from a particular population (see Edgington, 1966; Edgington & Onghena, 2007; Shadish, Cook, & Campbell, 2002). This involves drawing inferences about populations and settings that seem similar to the participants and settings involved in the study conducted, especially in terms of characteristics that appear relevant. For example, a researcher might make nonstatistical inferences about students that have similar educational and socioeconomic backgrounds as those in our study.

6.6 SUMMARIZING THE RESULTS FOR PUBLICATION

When reporting the results from a randomization or permutation test, it is important to report the method used to analyze the data along with the p-value. It is also necessary to report whether the p-value is exact or approximated via Monte Carlo methods. If the latter is the case, the number of data permutations that were carried out should be reported, and whether the correction for the p-value was used. The following is an example that might be used to summarize the results from the after-school program research.

Sample Write-Up

Three-hundred fifty-six middle-school students were randomly assigned to either fully participate in an after-school program ($n = 169$), or were given 'treatment as usual' ($n = 187$). The treatment as usual students were invited to attend one after-school activity per month. A randomization test was used to determine whether there was a statistically reliable difference in the effect of delinquency between students in these two groups. A Monte Carlo p-value was computed by permuting the data 4999 times. Using a correction suggested by Davison and Hinkley (1997), a p-value of 0.116 was obtained. This is weak evidence against the null hypothesis of no treatment effect, and may suggest that after-school programs do not contribute to differences in delinquency between students who participate fully and those who do not participate.

6.7 EXTENSION: TESTS OF THE VARIANCE

In the educational and behavioral sciences, it is often the case that the researcher will test whether the locations of two distributions (i.e., means, medians) are equivalent. Sometimes it can also be more informative to examine whether the variation between two distributions is equivalent. For example, in research related to educational curriculum, comparisons are often made between old and new curricula. It is common for researchers to find no differences in the average achievement between old and new curricula, yet have the achievement scores of the new curricula be smaller than those of the old curricula. This is an example of a treatment effect that manifests itself through changes in variation rather than location. Again, consider the after-school program research. The original null hypothesis was that

H_0 : The after-school program does not have an effect on delinquency.

It was found that in terms of average levels of delinquency, the evidence was weak in refuting the null hypothesis. What about in terms of variation? The two samples show differences in variation, with the control group ($s^2 = 110.7$) having more variation in delinquency measures than the treatment group ($s^2 = 80.5$). Does the sample difference of 30.2 provide convincing evidence that the after-school program has an effect on the variation of delinquency measures? Specifically, the null hypothesis,

$$\sigma^2_{\text{Control}} = \sigma^2_{\text{Treatment}},$$

is tested. This hypothesis is equivalent to

$$\sigma^2_{\text{Control}} - \sigma^2_{\text{Treatment}} = 0.$$

The same process of simulation introduced earlier in the chapter can be used to obtain a Monte Carlo p-value for this analysis. The only difference is that instead of writing a function to compute the mean difference, the function needs to compute the difference in variances. Command Snippet 6.10 shows the syntax for carrying out 4999 permutations of the data and computing the difference in variances. Using the correction for Monte Carlo p-values,

$$p = \frac{1198 + 1}{4999 + 1} = 0.240. \tag{6.4}$$

The probability of obtaining a sample difference in variance as extreme or more extreme than the observed difference of 30.2 is 0.240. This p-value is weak evidence against the null hypothesis and does not indicate that there are differences in variation that are attributable to participation in after-school programs.

Command Snippet 6.10: Syntax to permute the difference in variances.

```
## Function to compute the difference in variances
> var.diff <- function(data) {
  var(data[1:187]) - var(data[188:356])
  }

## Carry out the var.diff() function on the original permuted
    data
> var.diffs <- apply(X = permuted, MARGIN = 2, var.diff)

## Compute the p-value
> length(var.diffs[abs(var.diffs) >= 30.2])
[1] 1198
```

6.8 FURTHER READING

Major sources on the theory of randomization include Kempthorne (1955, 1977) and Fisher (1935). Ernst (2004) provides a readable account of both randomization and permutation methods, including application of such procedures.   Edgington and Onghena (2007) provide an extensive treatise on randomization tests for many situations.   An implementation of permutation tests under a unified framework in R is explained in detail in Hothorn, Hornik, van de Wiel, and Zeileis (2008).

PROBLEMS

6.1   Use the data in the *AfterSchool.csv* data set to examine whether there are treatment effects of the after-school program on *victimization* measures.

- Carry out an exploratory analysis to initially examine whether there are treatment effects of the after-school program on victimization measures.

- Use the randomization test to evaluate whether there is convincing evidence that the after-school program has an effect on victimization.

Write up the results from both sets of analyses as if you were writing a manuscript for publication in a journal in your substantive area.

6.2   Using the data in the *AfterSchool.csv* data set, investigate how the number of permutations of the data may effect the statistical conclusion or inference regarding the treatment effects of the after-school program on victimization measures. Carry out the randomization test from the previous problem again, except this time permute the data (a)100; (b) 500; (c) 1000; (d) 5000; (e) 10,000; and (f) 100,000 times.

- **a)** Compare the *p*-value across each of these analyses and comment on how the number of permutations might effect the statistical conclusions. What advice would you offer other researchers who are beginning to use the randomization test in their own studies?

6.3   Read the methods section of the article by Gottfredson and her colleagues (Gottfredson et al., 2010), paying particular attention to the sample used in the study. (This article can be accessed online through the library.) Given what you know about the characteristics of the sample and the sampling design, describe the population that the results of the after-school program intervention could be generalized to.

CHAPTER 7

BOOTSTRAP TESTS

Left to our own devices, . . . we are all too good at picking out non-existent patterns that happen to suit our purposes.

—B. Efron & R. Tibshirani (1993)

In Chapter 6, the use of the randomization test as a method that a researcher can use to examine group differences was introduced under the auspices of random assignment. When using the randomization test to draw inferences regarding the observed data, the distribution of the mean difference was considered based on the potential random assignments that could have occurred. This distribution was referred to as the randomization distribution of the mean difference. By examining the expected variation in the randomization distribution, that is, the variation of the mean difference, a decision regarding the likelihood of the observed difference being due only to the random assignment could be made.

The key question addressed by using any statistical method of inference is "how much variation is expected in a particular test statistic simply because of the randomness employed?" With randomization tests, the "randomness employed" was random assignment. In this chapter the use of *bootstrap* methods to test for group differences is examined. Bootstrapping does not assume random assignment to groups, but rather

Comparing Groups: Randomization and Bootstrap Methods Using R **139**
First Edition. By Andrew S. Zieffler, Jeffrey R. Harring, & Jeffrey D. Long
Copyright © 2011 John Wiley & Sons, Inc.

that the samples were randomly drawn from some larger population. Here the key question is "how much variation would be expected in a particular test statistic, if one repeatedly draws *random samples* from the same population?"

Bradley Efron introduced the methodology of bootstrapping in 1979 as a computer-based simulation framework to replace the inaccurate and complicated approximations that theoretical methods provide.[1] The bootstrap methodology uses Monte Carlo simulation to resample many *replicate data sets* from a probability model assumed to underlie the population, or from a model that can be estimated from the data. The replicate data sets can then be used to examine and quantify the variation in a particular test statistic of interest. Moreover, depending on the information one has about the population model, the bootstrapping can be *parametric, semiparametric,* or *nonparametric.* In this chapter, the use of the *parametric bootstrap* and the *nonparametric bootstrap* as inferential methods to test for differences between distributions is discussed.

7.1 EDUCATIONAL ACHIEVEMENT OF LATINO IMMIGRANTS

Recall that in Chapter 6 two research questions were laid out. The first question regarding whether there is a difference in the effect of delinquency between students who participated in the after-school program and students that did not was answered in Chapter 6 using the randomization test. The second research question posed was whether there is a difference in the educational achievement of immigrants from Mexico and that of immigrants from other Latin American countries. This question came out of research being performed by Stamps and Bohon (2006) who studied the educational achievement of Latino immigrants by examining a random sample of the 2000 decennial Census data (a subset of which is provided in LatinoEd.csv). In the course of their research, they began to wonder whether there could be a link between where the immigrants originated and their subsequent educational achievement. This chapter explores whether there is a difference in the educational achievement of immigrants from Mexico and that of immigrants from other Latin American countries. The focus here is on statistical inference. Thus, the exploration entails exploring whether there are population differences among the groups.

After examining the codebook for the data and reading the data into R, an initial inspection of both graphical and numerical summaries suggest that there is a sample difference in educational achievement. Command Snippet 7.1 shows the syntax for the exploration of these data.

Figures 7.1 and 7.2 show that the distribution of educational achievement for both groups is roughly symmetric. For Latin American immigrants from Mexico ($M = 59$), the typical educational achievement score is approximately 6 points lower

---

[1]The nomenclature of bootstrapping comes from the idea that the use of the observed data to generate more data is akin to a method used by Baron Munchausen, a literary character, after falling "in a hole nine fathoms under the grass, ... observed that I had on a pair of boots with exceptionally sturdy straps. Grasping them firmly, I pulled with all my might. Soon I had hoist myself to the top and stepped out on terra firma without further ado" (Raspe, 1786./1948, p. 22).

(within rounding) than for those immigrants from other Latin American countries ($M = 66$). The achievement scores for Mexican immigrants also show more variation (SD $= 16$ vs. SD $= 13$) indicating that Latinos from countries other than Mexico are more homogenous in terms of their educational achievement level.

Command Snippet 7.1: Syntax to read in and examine the Latino education data.

```
## Read in the data
> latino <- read.table(file = "/Documents/Data/LatinoEd.csv",
    header = TRUE, sep = ",", row.names = "ID")

## Examine the data frame object
## Output is suppressed
> head(latino)
> tail(latino)
> str(latino)
> summary(latino)

## Density plots conditioned on Mex
> plot(density(latino$Achieve[latino$Mex == 0], bw = 5.5),
    main= " ", xlab = "Educational Achievement", bty = "l", lty
    = "dashed", xlim = c(0, 100))
> lines(density(latino$Achieve[latino$Mex == 1], bw = 5.5), lty
    = "solid")
> legend(x = 5, y = 0.030, legend = c("Non-Mexican Immigrants",
    "Mexican Immigrants"), lty = c("dashed", "solid"), bty =
    "n")

## Side-by-side box-and-whiskers plots
> boxplot(latino$Achieve[latino$Mex == 0],
    latino$Achieve[latino$Mex == 1], horizontal = TRUE)

## Conditional means
> tapply(X = latino$Achieve, INDEX = latino$Mex, FUN = mean)
        0       1
64.5147 58.5931

## Conditional standard deviations
> tapply(X = latino$Achieve, INDEX = latino$Mex, FUN = sd)
         0        1
13.03141 15.62688

## Sample sizes
> table(latino$Mex)

   0   1
  34 116
```

Similar to the question asked in Chapter 6, the key statistical question to be answered here is whether the difference of six achievement points is expected given the variation that one would expect in the mean difference just due to chance alone. Here, the chance variation is not due to differences in random assignment—after all,

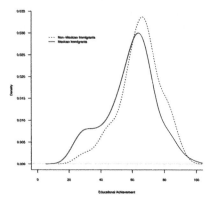

Figure 7.1: Density plots for the distribution of educational achievement conditioned on whether or not the Latin American country of emmigration was Mexico.

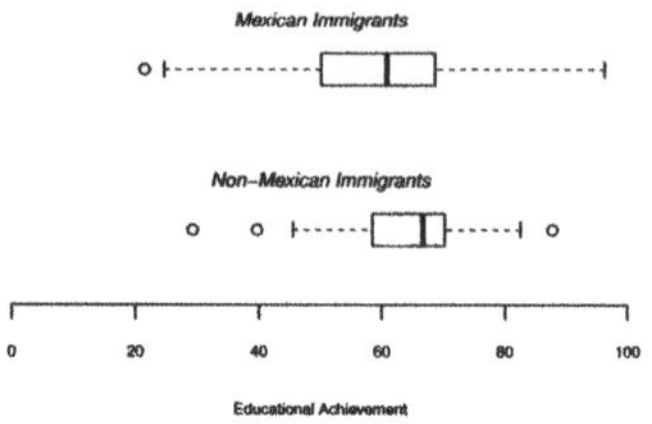

Figure 7.2: Side-by-side box-and-whiskers plots for the distribution of educational achievement conditioned on whether or not the Latin American country of emmigration was Mexico.

the two groups in this example were not randomly assigned—but rather it is due to random sampling. If a different random sample of Mexican and non-Mexican immigrants had been drawn, it would be expected that their mean achievement would be slightly different. This is likely, even when the population means for Mexican and non-Mexican Latino immigrants are identical.

How much of a difference in means is expected when drawing different random samples? Suppose it was possible to draw many different random samples from the populations of Mexican and non-Mexican immigrants. Then the extent of variation due to random sampling could be determined. This information could be used as a reference for evaluating the mean difference of six achievement points that were found for the single random sample considered above. If the probability model underlying the population(s) is known, or some assumptions about the probability model can be made, the *parametric bootstrap* can be used to simulate the distribution of the test statistic. Before further explanation of this methodology, a short digression is offered on the use of probability models in statistics—an essential element of the parametric bootstrap test.

7.2 PROBABILITY MODELS: AN INTERLUDE

The fundamentals of probability distributions can be illustrated with some coin flipping problems. Suppose a person tosses 10 fair coins and counts the number of heads. What are the possible outcomes? Now imagine the person tossing these 10 coins over and over again, and recording the number of heads each time. What is the probability attached to each of the possible outcomes?

Suppose the number of heads is denoted by X. Because the number of heads will vary each time the 10 coins are tossed, the variable X is referred to as a *random variable*. The possible outcomes of X comprise what is known as the *sample space*, which for this example can be mathematically expressed as

$$X = \{1, 2, 3, 4, 5, 6, 7, 8, 9, 10\}.$$

To be clear, these are all the possible number of heads that can occur when 10 coins are tossed. While there are 10 outcomes possible, they do not occur with equal probability. For example, there are multiple ways to get 5 heads,

H, H, T, T, H, T, T, T, H, H   or

H, T, H, H, H, T, H, T, T, T,

but only one way to get 10 heads, namely

H, H, H, H, H, H, H, H, H, H.

To find the probability attached to each of the possible outcomes, all the possible outcomes have to be enumerated. Once all of the outcomes have been enumerated, a probability can be assigned to each of the k possible outcomes. Each probability is symbolized as

$$P(X = k) = p_k \tag{7.1}$$

The probabilities have the constraints that,

- The individual probabilities must sum to one, $\sum p_k = 1$; and

- Each individual probability must be between the values of 0 and 1 inclusive, $0 \leq p_k \leq 1$.

A correspondence between each and every outcome and the associated probability is called a *probability distribution*. Probability distributions can be represented in a table, graph, or formula. For example, the probability distribution of X, the number of heads in 10 coin tosses, is presented in both Table 7.1 and Figure 7.3

7.3 THEORETICAL PROBABILITY MODELS IN R

How does one obtain probability values for the possible outcomes? In practice, it is common to specify a theoretical *probability distribution* that can be used as a *model* to assign probabilities for each outcome. These probability distributions are models because they don't exactly reproduce reality in terms of the populations they represent. To be useful, the specific probability assignments need to be consistent

Table 7.1: Probability Distribution of Random Variable X for Number of Heads When Tossing 10 Coins

Outcome	Probability	Symbolic Notation
0	0.001	$P(X = 0) = 0.001$
1	0.010	$P(X = 1) = 0.010$
2	0.044	$P(X = 2) = 0.044$
3	0.117	$P(X = 3) = 0.117$
4	0.205	$P(X = 4) = 0.205$
5	0.246	$P(X = 5) = 0.246$
6	0.205	$P(X = 6) = 0.205$
7	0.117	$P(X = 7) = 0.117$
8	0.044	$P(X = 8) = 0.044$
9	0.010	$P(X = 9) = 0.010$
10	0.001	$P(X = 10) = 0.001$

with reality. There are several probability distributions that serve as useful models in the educational and behavioral sciences. For example, tossing a coin 10 times can be modeled using the *binomial probability distribution*. The binomial distribution results from a process in which

- There are a fixed number n of observations.

- Each observation is independent.

- Each observation results in one of only two possible outcomes—called a *success* or *failure*.

- The probability of a success is the same for each observation.

Our coin tossing example is an example of a binomial process, because each of the 10 tosses is independent of one another. Each toss results in one of two possible outcomes—heads or tails, and the probability of a head (success) is 0.5 for each toss, assuming a fair coin.[2] R has several built-in probability distributions in addition to the binomial that are useful in modeling "real-world" processes or populations. A list of probability distributions included in base R, as well as other probability distributions available in add-on packages can be found at http://cran.r-project.org/web/views/Distributions.html.

---

[2]Sometimes this is referred to as a Bernoulli process rather than a binomial process.

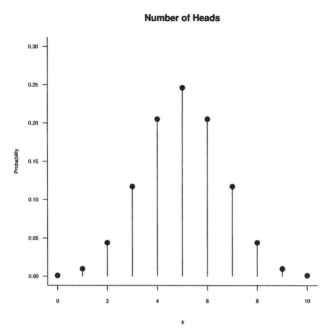

Figure 7.3: Spike plot of the probability distribution of the outcome for the number of heads in 10 tossed coins.

In parametric bootstrapping, observations are sampled—with replacement—from a particular probability distribution (i.e., population). To draw random values from a probability distribution, the prefix r is attached to the root name of the distribution. For example, to draw random observations from a population or process that is normally distributed, the r prefix is attached to the root name norm. This results in a function rnorm(). The website mentioned above also provides the root names for all the probability distributions in the base R system.

7.4 PARAMETRIC BOOTSTRAP TESTS

In the Latino data, the observed sample mean difference is 5.82 achievement points. Is this enough evidence to suggest that there are population differences in educational achievement between Mexican and non-Mexican Latino immigrants? To examine this question, the following null hypothesis is tested:

H_0 :There is no population difference in achievement scores
between Mexican and non-Mexican immigrants.

Similar to what was done in Chapter 6, the null hypothesis is assumed to be true and then the variation in the distribution of the test statistic of interest (e.g., mean difference) is examined under this assumption. In this case, the null hypothesis is a statement about the population of achievement scores for Mexican and non-Mexican immigrants. In particular, it claims that the underlying probability distributions for both populations are identical. This is equivalent to the claim that both samples were randomly drawn from a single probability distribution. This probability distribution for the population of achievement scores is specified (i.e., a parametric model is chosen), and the random sampling that took place in the study is replicated by resampling from this distribution.

7.4.1 Choosing a Probability Model

How does one choose a particular probability model to use in our parametric boot-strap? This is a very good question to ask, and a very difficult question to answer. The model needs to be chosen to reflect the probability distribution of the population that underlies the observations from the original sample. The decision of which model to sample from can have drastic effects on the results and thus, on the inferences that one draws.[3] Because of this, the choice of model should be *guided by theoretical or substantive knowledge*.

Lest the bootstrap methodology is boo-hooed because of the seemingly subjective choice of the underlying model, it is important to point out that the choice of parametric models is not limited to bootstrap methods. Most inferential methods employed in statistics make some assumption about the probability distribution of the underlying population. For example, the independent samples *t*-test—a commonly used method used to test mean differences between two populations—makes the assumption that the probability distribution of scores underlying both populations is normal. This implies that one can do no worse with the bootstrap if normality is assumed.

7.4.2 Standardizing the Distribution of Achievement Scores

Prior empirical research on achievement suggests that the normal distribution is a reasonable choice for our parametric model. The normal distribution, like almost all probability distributions, is really a family of distributions. The family is defined by the mean and the standard distribution. The distribution will always be bell-shaped, but can be located at different mean values, and can be skinnier or wider depending on the value of the standard deviation. Since there are many normal distributions, there is a need to further define the distribution by specifying the mean and standard deviation. This is referred to as *parameterizing* the distribution.

This is an even more difficult task than choosing the probability distribution, since the researcher typically has no idea what the mean and standard deviation of the population of scores should be. The sample data could, of course, be used to estimate

---

[3]Because the validity of the inferences drawn is tied to the choice of parametric model, some researchers have proposed the examination of several models to examine the robustness of the results.

these parameters, however, these estimates are often biased. Luckily, this can be easily side-stepped by re-sampling from the *standard normal distribution*.

The standard normal distribution is parameterized having a mean of zero, and a standard deviation of one. Symbolically this distribution is expressed as $\mathcal{N}$(mean = 0, standard deviation = 1), or more simply, $\mathcal{N}(0, 1)$. To express that a variable has this normal distribution,

$$\sim \mathcal{N}(0, 1),\qquad(7.2)$$

is used. While this distribution is completely parameterized, there is still the problem that the original achievement observations are not on this new scale. In order to transform the original achievement scores to this scale, a common method called *standardizing* is used to reexpress the data on this scale. To standardize data, the mean score is subtracted from each of the observed scores and then this difference is divided by the standard deviation. This reexpresses the data using a metric that quantifies each observation in terms of its distance from the mean expressed in standard deviation units. The transformed scores are called z-scores. Mathematically they are computed as

$$z_i = \frac{y_i - \bar{y}}{\text{SD}_y}.\qquad(7.3)$$

A vector of z-scores for the Latino achievement scores is created either by carrying out the computations directly or by using the scale() function. This function takes the argument x= which is used to define the vector of scores to standardize. In Command Snippet 7.2 the scale() function is used to standardize the Achieve variable using the marginal mean and standard deviation. This new vector of standardized scores is also appended into the latino data frame.

Figures 7.4 and 7.5 show the conditional density plots and side-by-side box-and-whiskers plots for the standardized achievement scores. The shapes of these distributions have not changed after standardization.[4] The location of these distributions have shifted. The mean difference for the standardized achievement scores between the Mexican ($M = -0.09$) and non-Mexican ($M = 0.30$) immigrants is 0.39. Recall that the variation has also changed so that these scores are expressed in standard deviation units. Thus, the mean difference of 0.39 suggests that the non-Mexican immigrants, on average, score 4/10 of a standard deviation higher than the Mexican immigrants. In the next section the use of the parametric bootstrap to test whether this sample difference provides statistical evidence that Mexican and non-Mexican immigrants have different mean achievement in the population is explained.

---

[4]It is a common misconception that standardizing normalizes a distribution. This is probably due to tests that use z-scores from a normal distribution. It is stressed that computing z-scores based on sample data does not change the original distribution.

Command Snippet 7.2: Creating a vector of standardized achievement scores.

```
## Standardize the achievement scores
> latino$z.achieve <- scale(x = latino$Achieve)

## Re-examine the latino data frame
> head(latino)
  Achieve ImmYear ImmAge English Mex    z.achieve
1    59.2    77.7     9.6       1   1  -0.04824848
2    63.7    65.8     1.1       1   1   0.24701645
3    62.4    63.6     6.1       0   1   0.16171769
4    46.8    55.3     2.1       1   1  -0.86186738
5    67.6    73.1     2.3       1   1   0.50291272
6    63.1    75.7     8.4       1   0   0.20764779

## The marginal mean of the standardized scores is 0
> mean(latino$z.achieve)
[1] 6.27847e-17

## The marginal standard deviation of the standardized scores
     is 1
> sd(latino$z.achieve)
[1] 1

## The conditional means
> tapply(X = latino$z.achieve, INDEX = latino$Mex, FUN = mean)
           0           1
  0.30047291 -0.08806965
```

7.5 THE PARAMETRIC BOOTSTRAP

In this section, an explanation of the methodology underlying the parametric bootstrap is provided. Consider a sample of data, $y_1, y_2, y_3, \ldots y_n$ drawn from an infinitely large population that has a known probability distribution. The steps to bootstrap the distribution for a particular test statistic, $\hat{V}$, are provided in Figure 7.6.

7.5.1 The Parametric Bootstrap: Approximating the Distribution of the Mean Difference

To perform a parametric bootstrap under the null hypothesis that there are no differences in the educational achievement scores between Mexican and non-Mexican Latino immigrants, a parametric model to bootstrap from is specified, the standard normal distribution. Two replicate samples are then bootstrapped from this population distribution—one having $n_1 = 34$ observations and the other having $n_2 = 116$ observations, as these are the original group sample sizes. The mean difference between these two replicate samples is then computed. This process is repeated many times, say $R = 4999$.

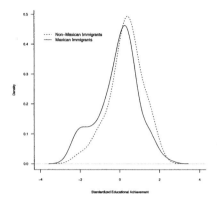

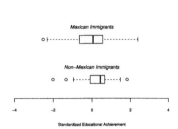

Figure 7.4: Density plots for the distribution of standardized educational achievement conditioned on whether or not the Latin American country of emmigration was Mexico.

Figure 7.5: Side-by-side box-and-whiskers plots for the distribution of standardized educational achievement conditioned on whether or not the Latin American country of emmigration was Mexico.

Because the null hypothesis is assumed to be true, both replicate samples are sampled from the same probability distribution. This computationally simplifies the process in that one need only draw a single vector of size $n = 34 + 116 = 150$ from the standard normal distribution and then compute the mean difference using the first 34 and the last 116 observations. (Or the first 116 and the last 34—since this is a random draw from the population, which observations are put in which group is really a moot point.) Thus, the key steps in the simulation, which is illustrated in Figure 7.7, are

- Bootstrap 150 observations from a standard normal distribution.

- Compute the mean difference between the first 34 and the last 116 observations.

- Repeat this process many times, each time recording the bootstrapped mean difference.

In Chapter 6, the `replicate()` and `apply()` functions were used to repeat computations such as permutations. These same functions could also be used to produce many bootstrap replicate data sets and obtain the bootstrap distribution of the mean difference. The **boot** package in R, however, is a much more elegant and complete implementation of the bootstrap methodology. This package comes as part of the base R system and only needs to be loaded into the session using the `library()` function.

Parametric Bootstrapping

- Randomly generate n observations $\{y_1^*, y_1^*, y_2^*, y_3^*, \ldots, y_n^*\}$ from the population distribution. This is called a *bootstrap replicate* because it is a replication of the observed data in the sense that it is another possible sample of size n that could have been drawn from the known population.

- Compute $\hat{V}$—the test statistic of interest—using the newly generated observations $y_1^*, y_1^*, y_2^*, y_3^*, \ldots, y_n^*$.

- Repeat these first two steps many times, say R times, each time recording the statistic $\hat{V}$.

- The distribution of $\hat{V}_1, \hat{V}_2, \hat{V}_3, \ldots, \hat{V}_R$ can be used as an estimate of the exact sampling distribution of V. This is called the *bootstrap distribution of V*.

Figure 7.6: Steps for bootstrapping the distribution for a particular test statistic, $\hat{V}$, from a known probability distribution.

7.6 IMPLEMENTING THE PARAMETRIC BOOTSTRAP IN R

The boot() function—from the **boot** package—can be used to generate bootstrap replicates of any statistic(s) of interest. To perform a parametric bootstrap, this function takes four required arguments. The first argument, data=, specifies the name of the data frame or vector of scores constituting the observed data. The argument R= provides the number of bootstrap replications to be performed.

The arguments sim="parametric" and ran.gen= each require a user-written function—the first describes how the bootstrap observations will be randomly sampled (e.g., which probability distribution, etc.), and the second indicates how the test statistic will be computed using the bootstrapped resamples.

7.6.1 Writing a Function to Randomly Generate Data for the boot() Function

For use in the boot() function, the random data generation function requires two arguments, data and mle. The first argument, data, takes the data frame or vector of scores provided in the boot() function in the random data generation (e.g., to compute how many observations to sample). The second argument, mle, consists of any other information needed for the process of data generation (e.g., parameter estimates).

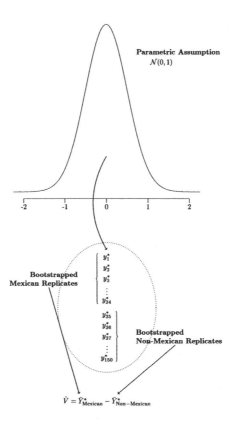

Figure 7.7: Visualization of the process used to parametrically bootstrap 150 observations from the standard normal distribution. The first 34 observations, the X replicates, constitute the replicated "non-Mexican immigrants" and the last 116 observations, the Y replicates, constitute the replicated "Mexican immigrants." $\hat{V}$ is the test statistic (e.g., the mean difference) computed for each replicate data set.

To write a function to randomly sample observations from the standard normal probability distribution, the `rnorm()` function is used. This function takes the arguments `n=`, `mean=`, and `sd=`. The first argument indicates how many observations to randomly draw, and the latter two arguments specify the parameters that define the probability distribution from which to draw. To draw from the standard normal distribution, for example, the arguments `mean=0` and `sd=1` are used. Note that these values are the default values for the arguments, so they do not need to be explicitly included in this example.

The first argument can be a particular number (e.g., n=150), but is instead provided the expression `length(data)`. By using this expression rather than an actual value, the function will sample the same number of observations that are included in the original vector of scores assigned to the object `data`, which is specified in the `boot()` function. In Command Snippet 7.3 a function, called `lat.gen()`, is written to randomly sample observations from the standard normal probability distribution.

Command Snippet 7.3: A function to sample 150 observations from the standard normal distribution. Although the argument `mle` is not used in this function, it needs to be included in the function for the `boot()` function to work.

```
> lat.gen <- function(data, mle){
      rnorm(n = length(data), mean = 0, sd = 1)
      }
```

7.6.2 Writing a Function to Compute a Test Statistic Using the Randomly Generated Data

A function that will compute a test statistic—in this example, the mean difference—for each of the bootstrapped replicate data sets is also needed. This function requires that the argument, `data`, be specified (similar to the function written in Chapter 6 to compute the mean difference) and an additional argument, `indices`. The first argument obtains the data randomly generated in the function `lat.gen()`. The second argument is a vector of indices, frequencies, or weights defining the observations from the bootstrap sample to be used.[5] Similar to the `mle` argument, in the random data generation function, this argument is included, but it will not be used by the function during this computation. In Command Snippet 7.4, a function, `mean.diff()`, is written that will compute the mean difference between the first 34 bootstrapped observations and the last 116 bootstrapped observations.

Command Snippet 7.4: A function to compute the mean difference in the bootstrapped achievement scores.

```
> mean.diff <- function(data, indices) {
  mean(data[1:34]) - mean(data[35:150])
  }
```

---

[5]One can improve the computational efficiency by only using particular observations or weighting certain observations as more important. This is beyond the scope of this monograph, but Kahn (1954) provides a fairly readable treatment of these ideas.

Both the `lat.gen()` and `mean.diff()` functions can be executed so that they can be implemented in the `boot()` function. Command Snippet 7.5 shows the final implementation of the parametric bootstrap via the `boot()` function. The two functions `lat.gen()` and `mean.diff()` are provided to the arguments `ran.gen=` and `statistic=`, respectively. Note that the parentheses are not needed when including the functions in the `boot()` function. The argument `R=4999` is also included to draw 4999 bootstrap replicates. The results are then assigned to an object `par.boot`.

Command Snippet 7.5: The use of the `boot()` function to perform a parametric bootstrap.

```
## Load the boot package
> library(boot)

## Carry out the parametric bootstrap
> par.boot <- boot(data = latino$z.achieve, statistic =
    mean.diff, R = 4999, sim = "parametric", ran.gen = lat.gen)
```

7.6.3   The Bootstrap Distribution of the Mean Difference

The object `par.boot` contains several elements including the 4999 bootstrapped mean differences. To list the components of the bootstrap object that are accessible, the `str()` function is called on the assigned bootstrap object (see Command Snippet 7.6). Each of the 4999 bootstrap mean differences is contained in the `t` element of the bootstrap object `par.boot`.

Similar to the process used in Chapter 6 to examine the permutation distribution, the bootstrap distribution of the standardized mean difference is plotted. Command Snippet 7.7 shows the syntax to plot the 4999 replicate mean differences shown in Figure 7.8. This distribution is an approximation of the exact sampling distribution of the mean difference under the normal distribution. When the number of bootstrap replicates, R, is infinite ($R \to \infty$) then the bootstrap distribution and the exact sampling distribution are identical. In practice, however, the number of bootstrap replicates drawn is finite, so the bootstrap distribution is only an approximation of the exact distribution. This approximation differs only because of simulation error.

Note that the bootstrap distribution of the sample mean difference is roughly normally distributed. It is also roughly centered around zero—the difference between the population means assumed under the null hypothesis. This is confirmed by examining the mean of the bootstrap distribution, which is approximately zero. This suggests that the average mean difference, under the null hypothesis, is zero. The standard deviation of the bootstrap distribution is the approximation of the standard error (SE) for the mean difference. The SE of 0.2 suggests that there is not a lot of variation from the average mean difference of zero. The mean differences do not vary much just because of the random sampling.

Command Snippet 7.6: Listing the elements from the bootstrap object.

```
> str(par.boot)
List of 10
 $ t0        : num -0.243
 $ t         : num [1:4999, 1] -0.334 -0.193 -0.181 0.114 0.219
   ...
 $ R         : num 4999
 $ data      : num [1:150, 1] -0.0482 0.247 0.1617 -0.8619
      0.5029 ...
  ..- attr(*, "scaled:center")= num 59.9
  ..- attr(*, "scaled:scale")= num 15.2
 $ seed      : int [1:626] 403 426 -1087594034 -1689997951
      -1154526649 800955270 -1940498415 1227735036 -619234920
      -9902028 ...
 $ statistic:function (data, indices)
  ..- attr(*, "source")= chr [1:3] "function(data, indices) {"
    ...
 $ sim       : chr "parametric"
 $ call      : language boot(data = latino$z.achieve, statistic
      = mean.diff, R = 4999,       sim = "parametric", ran.gen =
      lat.gen)
 $ ran.gen   :function (data, mle)
  ..- attr(*, "source")= chr [1:3] "function(data, mle){" ...
 $ mle       : NULL
 - attr(*, "class")= chr "boot"
```

Command Snippet 7.7: Examine the bootstrap distribution of the mean difference.

```
## Plot the bootstrap distribution
> plot(density(par.boot$t))

## Draw a vertical line at 0
> abline(v=0)

## Mean of the bootstrap distribution
> mean(par.boot$t)
[1] 0.002673093

## Standard deviation of bootstrap distribution
> sd(par.boot$t)
[1] 0.1958675
```

Recall that the p-value measures the evidence against the null hypothesis by quantifying the probability of observing a sample statistic at least as extreme as the one estimated in the sample data assuming the null hypothesis is true. For the bootstrap test, this boils down to finding the proportion of times a bootstrapped statistic is at least as extreme as the observed statistic. In the example, the observed sample mean difference in standardized achievement scores is 0.39. Using the length() function and indexing, as shown in Command Snippet 7.8, the number of bootstrapped statistics that are *greater than or equal to* 0.39 and the number of bootstrapped statistics that are *less than or equal to* −0.39 are counted.

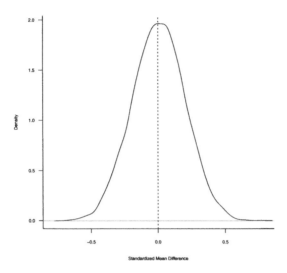

Figure 7.8: Bootstrap distribution of the mean difference in achievement scores between non-Mexican and Mexican immigrants for 4999 replicates under the hypothesis of no difference. The vertical line is drawn at 0.

Command Snippet 7.8: Syntax to count the number of bootstrapped statistics that are as extreme or more extreme than the observed mean difference.

```
>   length(par.boot$t[abs(par.boot$t) >= 0.39])
[1]  231
```

Using the Monte Carlo adjustment introduced in Chapter 6, this suggests that the probability of observing a sample mean difference as extreme or more extreme than the observed sample difference of 0.39—assuming that there is no population mean achievement score difference between Latino immigrants from Mexico and Latino immigrants from other countries—is

$$p = \frac{231 + 1}{4999 + 1}$$
$$= 0.046.$$

The *p*-value of 0.046 is moderate evidence against the null hypothesis.

7.7 SUMMARIZING THE RESULTS OF THE PARAMETRIC BOOTSTRAP TEST

The results of the parametric bootstrap test are reported similarly to other Monte Carlo inferential methods. Good (2005b, p. 232) suggests that when reporting the results, one should "provide enough detail that any interested reader can readily reproduce your results." Details would include the chosen test statistic, the assumptions of the method, and any results obtained. For example, the test of the mean educational achievement differences for the Latino data via the parametric bootstrap might be reported as follows.

> **Sample Write-Up**
>
> A parametric bootstrap test was used to determine whether there was a population difference in the mean educational achievement scores between immigrants from Mexico and immigrants from other Latino countries. The data, which were standardized prior to analysis, were a random sample, ensuring that the assumption of independence for the parametric bootstrap test was met. A Monte Carlo p-value was computed by drawing 4999 bootstrap replicates from an assumed probability model of $\mathcal{N}(0,1)$. Using a correction suggested by Davison and Hinkley (1997), a p-value of 0.046 was computed. This is moderate evidence against the null hypothesis of no population differences, and suggests that Mexican and non-Mexican immigrants may differ in their educational achievement.

7.8 NONPARAMETRIC BOOTSTRAP TESTS

When differences between the Mexican and non-Mexican achievement scores were tested using the parametric bootstrap, an assumption was made that the probability model underlying the population of achievement scores was known. What if there was no theoretical knowledge to suggest the particular probability model? The density plot of the standardized achievement scores is again examined, but this time with the addition of variability bands (see Section 3.4). Command Snippet 7.9 shows the syntax for plotting the kernel density estimates with variability bands for both the Mexican and non-Mexican immigrants using the sm.density() function.

Figures 7.9 and 7.10 show the resulting plots. These plots suggest very slight misfit between the normal probability model and the observed data for the Mexican achievement scores. Depending on the degree of misfit between the normal model and the data, a researcher may no longer believe that the population of achievement scores is normally distributed, or may want to examine the validity of the parametric results by seeing whether the level of evidence changes if one no longer imposes this particular probability model on the data. If there is no substantive knowledge to support the use of a different parametric model, the probability model for the population can remain completely unspecified. The empirical distribution (i.e., the distribution suggested by the sample data) can then be used as a proxy for the population distribution and the researcher can use *nonparametric bootstrapping* to test for population differences.

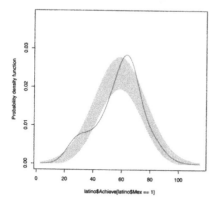

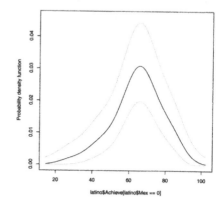

Figure 7.9: Density plot (dashed line) with variability bands of the distribution of standardized educational achievement for Mexican immigrants. The solid line is the density curve of the standard normal probability model.

Figure 7.10: Density plot (dashed line) with variability bands of the distribution of standardized educational achievement for non-Mexican immigrants. The solid line is the density curve of the standard normal probability model.

Command Snippet 7.9: Syntax to plot the kernel density estimates for both the Mexican and non-Mexican immigrants along with variability bands for the normal model.

```
## Load the sm library
> library(sm)

## Plot the kernel density estimate of achievement for
    non-Mexican immigrants
> sm.density(latino$Achieve[latino$Mex == 0], model = "normal",
    rugplot = FALSE)

## Plot the kernel density estimate of achievement for Mexican
    immigrants
> sm.density(latino$Achieve[latino$Mex == 1], model = "normal",
    rugplot = FALSE)
```

Nonparametric bootstrapping is another Monte Carlo simulation method for approximating the variation in the distribution of a test statistic. Nonparametric refers to the idea that the sample data are not generated from a particular parameterized probability model. In fact, nonparametric bootstrapping makes no assumptions about the probability model underlying the population. This method uses the distribution of the observed sample data, the *empirical distribution*, as a proxy for the population distribution. Replicate data sets are then randomly generated, with replacement from the empirical data. Since, in theory, the observed sample data should represent the population from which they are drawn, replicates drawn from these data should represent what one would get if many samples from the population were drawn. Figure 7.11 represents the concept of the nonparametric bootstrap.

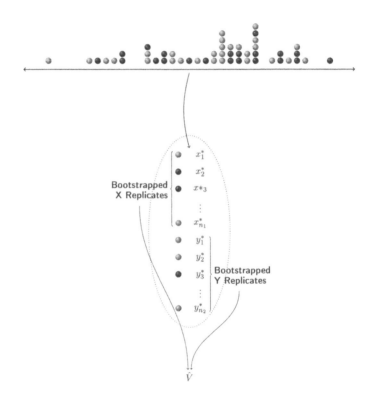

Figure 7.11: Nonparametric bootstrapping involves resampling from the pooled observed sample with replacement. The bootstrap distribution of a statistic obtained from this method is typically a good approximation of the exact distribution.

Consider a sample of data $\{y_1, y_2, y_3, \ldots y_n\}$ drawn from an infinitely large population that has some unknown probability distribution. Because the probability is unknown, the empirical distribution based on the observed data is used as a representation of the population distribution. The steps used to bootstrap the distribution for a particular test statistic, $\hat{V}$, are provided in Figure 7.12.

Nonparametric Bootstrapping

- Randomly resample n observations from the observed sample *with replacement* $\{y_1^*, y_1^*, y_2^*, y_3^*, \ldots, y_n^*\}$. This is called a *bootstrap replicate* because it is a replication of the observed data in the sense that it is another possible sample of size n that could have been drawn from the unknown population.

- Compute $\hat{V}$—the test statistic of interest—using the bootstrapped observations $\{y_1^*, y_1^*, y_2^*, y_3^*, \ldots, y_n^*\}$.

- Repeat these first two steps many times, say R times, each time recording the statistic $\hat{V}$.

- The distribution of $\hat{V}_1, \hat{V}_2, \hat{V}_3, \ldots, \hat{V}_R$ can be used as an estimate of the exact sampling distribution of $\hat{V}$. This is called the *bootstrap distribution of $\hat{V}$*.

Figure 7.12: Steps for using the sample data to bootstrap the distribution for a particular test statistic, $\hat{V}$, when the probability distribution is unknown.

7.8.1   Using the Nonparametric Bootstrap to Approximate the Distribution of the Mean Difference

Because the null hypothesis of no difference is again assumed to be true, the sample observations can once again be pooled together and replicate bootstrap samples from the pooled distribution can be drawn (see Figure 7.11). This again simplifies the process in that one sample of size $n = 150$ from the observed data can be drawn and the mean difference can be computed using the first 34 and the last 116 observations. Because there is no assumptions made about the population distribution of achievement scores, there is no longer a need to bootstrap the standardized scores. Instead, the raw achievement scores can be used in the nonparametric bootstrap. Consider the key steps involved in using a nonparametric bootstrap to obtain the distribution of the mean difference in educational achievement for the Latino data.

- Bootstrap from the 150 raw observations in the pooled sample.

- Compute the mean difference between the first 34 and the last 116 bootstrapped observations.

- Repeat this process many times, each time recording the bootstrapped mean difference.

7.8.2   Implementing the Nonparametric Bootstrap in R

The nonparametric bootstrap is also implemented in R using the boot() function. For nonparametric bootstrapping, only the arguments data=, statistic=, and R= are used. Because of the manner in which the boot() function resamples, in nonparametric bootstrapping, the data= argument takes the name of a data frame containing the observed data, even though one is really only interested in the vector of achievement scores. This is because the boot() function uses the row numbers of the data frame to keep track of which observations were resampled during the bootstrapping process. The argument R=, similarly to parametric bootstrapping, is the number of bootstrap replicates to be drawn. For nonparametric bootstrapping, the argument statistic= again takes a function that when applied to the bootstrapped data returns the statistic of interest. But, this time an expression is included in the function that indicates the computation should be carried out on the observations drawn in the bootstrap process.

Command Snippet 7.10 includes a function called mean.diff.np() that when applied to the Latino data will compute the mean difference between the first 34 resampled achievement scores and the last 116 resampled achievement scores. Note that the first line of this function uses indexing to create a new data frame object called d. The rows of d are the bootstrapped rows—which were stored internally in indices=—from the original data frame. This new data frame d is then used in all remaining computations in the function. The snippet also includes the syntax for carrying out the nonparametric bootstrap on the Latino data.

Command Snippet 7.10:   A function to compute the mean difference between the first 34 resampled observations and the last 116 resampled observations. The first line indicates that all rows of the data frame are to be used. The boot() function is then used to bootstrap the mean difference in achievement scores for 4999 replicates.

```
## Function to compute the mean difference
> mean.diff.np <- function(data, indices) {
  d <- data[indices, ]
  mean(d$Achieve[1:34]) - mean(d$Achieve[35:150])
  }

## Carry out the nonparametric bootstrap
> nonpar.boot <- boot(data = latino, statistic = mean.diff.np,
    R = 4999)
```

A plot of the bootstrap distribution is shown in Figure 7.13. Command Snippet 7.11 shows the syntax to examine the bootstrap distribution of the standardized mean difference. The distribution of these 4999 bootstrapped mean differences is again an approximation of the exact sampling distribution of the mean difference. Note that here again the bootstrap distribution of the sample mean difference is roughly normally distributed. It is also roughly centered around zero—the population mean difference assumed in the null hypothesis. These mean differences, again, show very little variation, having a small SE of approximately 2.9. Command Snippet 7.11 also shows the syntax to count the number of bootstrapped standardized mean differences that are as extreme or more extreme than the observed standardized mean difference of 0.39.

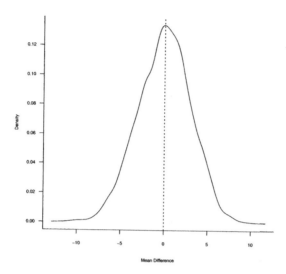

Figure 7.13: Bootstrap distribution of the mean difference (using nonparametric bootstrapping) in achievement scores between non-Mexican and Mexican immigrants for 4999 replicates assuming there are no population differences. The vertical line is drawn at zero.

Using the Monte Carlo adjustment, this suggests that the probability of observing a sample mean difference as extreme or more extreme than the observed sample difference of 5.9 standard deviations—assuming that there is no population differences in standardized achievement scores between Latino immigrants from Mexico and Latino immigrants from other countries—is

$$p = \frac{225 + 1}{4999 + 1}$$
$$= 0.045.$$

This simulated p-value of 0.045 is moderate evidence against the null hypothesis. It is noted that the p-value for the nonparametric bootstrap is very similar to that of the parametric bootstrap. This consistency speaks to the robustness of the result under different assumptions about the probability distributions.

Command Snippet 7.11: Syntax to examine the bootstrap distribution of the mean difference from the nonparametric bootstrap, and count the number that are as extreme or more extreme than the observed standardized mean difference of 0.39.

```
## Plot the bootstrap distribution
> plot(density(nonpar.boot$t))

## Draw a vertical line at 0
> abline(v=0)

## Mean of the bootstrap distribution
> mean(nonpar.boot$t)
[1] 0.06338843

## Standard error of bootstrap distribution
> sd(nonpar.boot$t)
[1] 2.942154

## Count the mean differences as or more extreme than 0.39
> length(nonpar.boot$t[abs(nonpar.boot$t) >= 5.9])
[1] 225
```

7.9 SUMMARIZING THE RESULTS FOR THE NONPARAMETRIC BOOTSTRAP TEST

The results of the nonparametric bootstrap test are reported almost identically to those of the parametric bootstrap tests. For example, the test of mean educational achievement differences for the Latino data via the nonparametric bootstrap might be reported as follows.

Sample Write-Up

A nonparametric bootstrap test was used to determine whether there was a population difference in the mean educational achievement scores between immigrants from Mexico and immigrants from other Latino countries. The data were a random sample, ensuring that the assumption of independence for the nonparametric bootstrap test was met. A Monte Carlo p-value was computed by drawing 4999 bootstrap replicates from the data. Using a correction suggested by Davison and Hinkley (1997), a p-value of 0.045 was computed. This is moderate evidence against the null hypothesis of no population differences, and suggests that Mexican and non-Mexican immigrants may differ in their educational achievement.

7.10   BOOTSTRAPPING USING A PIVOT STATISTIC

In the current example, the test statistic of interest is the mean difference. There are of course many different statistics one could choose. How does one choose a statistic? In order to answer this, the null hypothesis needs to be considered. The null hypothesis is really a statement that specifies some or all of the parameters in the assumed probability model. For example, the null hypothesis of no difference in the population distributions of achievement scores suggests that both the population distribution for the Mexican and non-Mexican achievement scores are identical.

A test statistic is used as a measure of the discrepancy of the sample with respect to the null hypothesis. For example, by choosing to examine the mean difference, the way the samples are discrepant with respect to the populations being identical is in their locations (i.e., means). Some researchers, in fact, write the null hypothesis in terms of the population parameters that they believe are discrepant. For example,

$$H_0 : \mu_{\text{Mexican}} = \mu_{\text{non-Mexican}}.$$

This hypothesis can also be written in terms of the hypothesized difference in location,

$$H_0 : \mu_{\text{Mexican}} - \mu_{\text{non-Mexican}} = 0.$$

Note that the difference in variances, or some other measure, could just as easily have been chosen as the test statistic, depending on how the difference in the samples was represented. The distribution of the test statistic is then built up from the assumption that the null hypothesis is true—if there is no difference, the average standardized mean difference from all possible random samples that could have been drawn is zero.

In the first example using the parametric bootstrap, the achievement scores were initially standardized to have a mean of zero and standard deviation of one. Recall that this was done so that the probability model used was completely parameterized. Otherwise, even with a specified probability model such as the normal distribution, there still would have been many normal models from which to choose (e.g., models where the standard deviation could vary).

When the probability model is not completely specified, the p-value is generally not well defined because of its dependence on which probability model is selected. For example, the p-value might have changed depending on the standard deviation that was selected for the normal probability model. In practice, there are different ways in which statisticians deal with this problem. One way to deal with the problem is to choose a test statistic that has the same distribution under the null hypothesis for any set of parameters that are chosen for a particular family of models. Such a statistic is referred to as a *pivot statistic*.

7.10.1   Student's *t*-Statistic

One such pivot statistic for examining the difference in location between two populations is *Student's t-statistic*. The *t*-statistic was developed by William Sealy Gosset whose work as a statistician at Guinness Brewery forced him to deal with small samples with around 8 to 10 observations.[6] At the time he published his landmark paper in 1908, there was only theoretical work in place to deal with sample sizes in excess of 150 observations. At that point in time, the standard method to examine the precision in a mean estimate was to examine the ratio of the mean to its standard error. As Senn (2008, p. 37) points out, "this created a problem, since the standard approaches to dealing with a precision of the mean related it to the standard error but made no allowance for the fact that in small samples the standard error itself was subject to considerable random variation." Gosset was able to investigate how the random variation impacted the precision of the mean estimate in small samples using empirical data. He then determined the sample sizes at which point Pearson's theoretical work could be applied, also providing alternative computations for sample sizes that were below these cutoffs.

Fisher was the first to acknowledge Gosset's work recognizing that the t-statistic was not dependent on "the unknown value of some nuisance parameter." He extended Gosset's initial example beyond the one-sample case, and also made it a major part of his classic book *Statistical Methods for Research Workers*. Fisher's adaptations to Gosset's original statistic allowed for testing two samples from different populations. The t-statistic can be computed as

$$t = \frac{\bar{X} - \bar{Y}}{\sqrt{s^2_{\text{pooled}} \left(\frac{1}{n_X} + \frac{1}{n_Y} \right)}}, \tag{7.4}$$

where $\bar{X}$ and $\bar{Y}$ are the mean for sample X and sample Y, n_X and n_Y are the sample sizes for sample X and sample Y, and s^2_{pooled} is the weighted average of the two sample variance estimates s^2_X and s^2_Y called the *pooled variance*. The pooled variance is computed as

$$s^2_{\text{pooled}} = \frac{(n_X - 1)s^2_X + (n_Y - 1)s^2_Y}{n_X + n_Y - 2}. \tag{7.5}$$

The numerator of the t-statistic is the mean difference. The denominator is the numeric approximation of the standard error of the mean difference. Note that this value is comparable to the standard error from the nonparametric bootstrap analysis presented earlier. The t-statistic is just the ratio of these two quantities which expresses the mean difference using the units of the standard error. Expressing observations using standard error units is referred to as *studentizing* the observations. Command Snippet 7.12 shows the computation of the studentized mean difference in achievement for the Latino data.

The observed studentized mean difference of 2.01 shows a difference of a little more than two standard errors between the mean achievement scores for Mexican and non-Mexican immigrants. Again the question is asked whether this observed

---

[6]Guiness had a strict no publication policy for their research scientists to deter their company's commercial secrets from being stolen. As such, Gosset was forced to publish under the psuedonym *Student*.

difference is large relative to the variation expected in the studentized mean difference due only to differences in random sampling. To examine this, the nonparametric bootstrap test is used. Command Snippet 7.13 shows the syntax to write a function called `studentized.mean.diff()` and also the application of this statistic in the `boot()` function. The adjusted Monte Carlo simulated p-value based on the t-statistic, $p = 0.044$, provides moderate evidence against the null hypothesis.

Command Snippet 7.12: A function to compute the studentized mean difference for the Latino achievement data.

```
## Compute the conditional sample means
> tapply(X = latino$Achieve, INDEX = latino$Mex, FUN = mean)
        0        1
  64.5147  58.5931

## Compute the conditional sample standard deviations
> tapply(X = latino$Achieve, INDEX = latino$Mex, FUN = var)
         0         1
  169.8177  244.1993

## Compute the sample sizes
> table(latino$Mex)

   0   1
  34 116

## Compute the mean difference
> numerator <- 64.5 - 58.6
> numerator
[1] 5.9

## Compute the pooled variance
> pool.var <- (33 * 169.8 + 115 * 244.2) / 148
> pool.var
[1] 227.6108

## Compute the standard error of the mean difference
> denominator <- sqrt(pool.var * (1 / 34 + 1 / 116))
> denominator
[1] 2.942210

## Compute the t-statistic
> t <- numerator/denominator
> t
[1] 2.005296
```

Table 7.2 shows the nonparametric bootstrap results using raw achievement scores, standardized achievement scores (not presented in the chapter), and the studentized scores. All three analyses produce comparable results based on the p-values. The major difference between using the standardized mean difference and the studentized mean difference is the time in the analysis at which the reexpression takes place. In the use of the standardized mean difference, the achievement scores are standardized prior to carrying out the bootstrap. This is referred to as *prepivoting* the data. In the use of the studentized mean difference, the bootstrapping is carried out on the raw data and the reexpression, or pivot, takes place in the computation of the test statistic. There is some evidence that prepivoting reduces the error in the rejection probability and increases the reliability of the inferences (see Beran, 1987; Efron, 1982; Hinkley & Wei, 1984).

Table 7.2: Results of Carrying Out Nonparametric Bootstrap Test for Three Different Test Statistics[a]

Statistic	Observed	SE	*p*-Value
Raw mean difference	5.92	2.94	0.045
Standardized mean difference	0.39	0.20	0.044
Studentized mean difference	2.01	1.02	0.044

[a] All tests used 4999 replications of the data.

7.11 INDEPENDENCE ASSUMPTION FOR THE BOOTSTRAP METHODS

As with any inferential method, the bootstrap test has certain assumptions that need to be met in order for valid inferences. The major assumption for both the parametric and nonparametric bootstrap test is that the observations in the original sample are *independent* of one another.

Observations are statistically independent if the value or occurrence of a particular measurement in no way influences or changes the probability of the occurrence or value of any other measurements. Another way to think about independence is that the measurements in one sample are unrelated (uncorrelated) with the measurements in the other sample—that is, the variation in one sample is unrelated to the variation in the other. Independence is a function of the data collection method. Random sampling ensures the sample observations are independent. In practice, the tenability of actually obtaining a random sample depends on the research context.

Without a random sample, justification for this assumption is incumbent on the researcher, and is often based on logical argument resting on the researchers knowledge of the substantive area. There is currently no statistical method that is available to directly assess whether the measurements in the two samples are independent. In research that does not employ random sampling, often the assumption of independence is violated because there are often correlations that exist among the study participants, such as genetic relationships or physical proximity. Such violations could result in a systematic underestimate of the variation in the population which leads to erroneous conclusions.

Command Snippet 7.13: A function to compute the *t*-statistic between the first 116 resampled observations and the last 34 resampled observations. The `boot()` function is then used to bootstrap the mean difference in achievement scores for 4999 replicates.

```
## Function to compute t-statistic
> studentized.mean.diff <- function(data, indices) {
  d <- data[indices, ]
  nonmex <- d$Achieve[1:34]
  mex <- d$Achieve[35:150]
  num <- mean(nonmex) - mean(mex)
  pool.var <- (33 * var(nonmex) + 115 * var(mex)) / 148
  den <- sqrt(pool.var * (1 / 34 + 1 / 116))
  num/den
  }

## Carry out nonparametric bootstrap
> studentized.boot <- boot(data = latino, statistic =
    studentized.mean.diff, R = 4999)

## Plot the bootstrap distribution
> plot(density(studentized.boot$t), xlab="Bootstrapped
    t-Statistic", main=" ")

## Mean of the bootstrap distribution
> mean(studentized.boot$t)
[1] 0.02357806

## Standard deviation of bootstrap distribution
> sd(studentized.boot$t)
[1] 1.016181

## Compute the p-value
> length(studentized.boot$t[abs(studentized.boot$t) >= 2.01])
[1] 219

## Compute the Monte Carlo adjusted p-value
> (219 + 1) / (4999 + 1)
[1] 0.044
```

There are two common situations in the educational and behavioral sciences, both related to the data collection method in which the independence assumption is generally violated. The first such violation occurs when there are multiple measurements made on the same participant. For example, a common research design in the educational and behavioral sciences is the pre–post design, in which measurements are taken from the same participants both prior to and after some intervention has occurred. The researcher using such designs is often interested in whether there are mean differences between the pre- and postintervention measurements. These two samples, which consist of measurements taken on the same individuals, would not be independent of one another.

Another common violation of independence, especially in educational research, occurs when the assignment of the intervention is at one level and the unit of analysis at a different level. For example, an educational researcher might assign an intervention at the classroom level (i.e., a whole classroom gets the intervention), and then mistakenly perform the analysis as though it were the students who were randomly assigned.

7.12   EXTENSION: TESTING FUNCTIONS

It is often good practice to test a function that has been written before using it in an analysis. For example, it would be good to know that the computations in the function written to compute the mean difference in Command Snippet 7.4 were being performed correctly. One method to test the function would be to execute it on the observed data, and then compare the results to the mean difference that was actually computed. Command Snippet 7.14 shows the results of executing `mean.diff()` on the vector of standardized achievement scores.

Command Snippet 7.14: Executing the function `mean.diff()` on the vector of standardized achievement scores.

```
> mean.diff(latino$z.achieve)
[1]  -0.2430629
```

The computed result from the `mean.diff()` function of −0.24 is not the same as the difference of 0.39 that was computed based on the conditional means using the `tapply()` function in Command Snippet 7.1. The problem, however, is not with the function.

The function computes the difference between the mean of the first 34 observations in the vector submitted to the function and the mean of the last 116 observations in the vector. The computed mean difference from the function is not the difference in the conditional means since the first 34 observations do not correspond to the 34 non-Mexican Latino immigrants in the data, nor do the last 116 observations correspond to the Mexican immigrants.

7.12.1   Ordering a Data Frame

In order to evaluate if the function is working properly, the data frame needs to be sorted so that the first 34 rows correspond to the non-Mexican immigrants and the last 116 rows correspond to the Mexican immigrants. This can be accomplished by ordering the rows based on the values of the `Mex` variable.

The `order()` function sorts a variable and produces output that can be used to sort the rows of a data frame. This function takes the name of the variable to be sorted on. Command Snippet 7.15 shows the results of ordering on the `Mex` variable. Indexing can then be used to order the data frame. Command Snippet 7.16 shows the syntax to order the rows of the `latino` data frame using the `Mex` variable.

Command Snippet 7.15: Syntax to return the order of the rows in the `latino` data frame based on the value of the `Mex` variable. The rows are in ascending order.

```
> order(latino$Mex)
  [1]    6   12   16   19   32   33   35   36   37   38   42   47   52
 [14]   56   71   76   78   79   80   82   84   86  102  105  110  116
 [27]  117  118  124  141  143  146  149  150    1    2    3    4    5
 [40]    7    8    9   10   11   13   14   15   17   18   20   21   22
 [53]   23   24   25   26   27   28   29   30   31   34   39   40   41
 [66]   43   44   45   46   48   49   50   51   53   54   55   57   58
 [79]   59   60   61   62   63   64   65   66   67   68   69   70   72
 [92]   73   74   75   77   81   83   85   87   88   89   90   91   92
[105]   93   94   95   96   97   98   99  100  101  103  104  106  107
[118]  108  109  111  112  113  114  115  119  120  121  122  123  125
[131]  126  127  128  129  130  131  132  133  134  135  136  137  138
[144]  139  140  142  144  145  147  148
```

Command Snippet 7.16: Syntax to order the rows in the `latino` data frame based on the value of the `Mex` variable. The rows are in ascending order.

```
## Order the rows of the latino data frame by Mex
> ordered.latino <- latino[order(latino$Mex), ]

## Examine the first part of ordered.latino
> head(ordered.latino)
     Achieve ImmYear ImmAge English Mex   z.achieve
6       63.1    75.7    8.4       1   0   0.20764779
12      46.2    81.6    8.7       1   0  -0.90123603
16      66.3    72.9    5.1       1   0   0.41761396
19      87.8    71.6    3.4       1   0   1.82832415
32      54.6    79.1    5.3       1   0  -0.35007484
33      59.2    72.0    4.6       1   0  -0.04824848

## Examine the last part of ordered.latino
> tail(ordered.latino)
     Achieve ImmYear ImmAge English Mex   z.achieve
140     83.9    81.4    8.6       1   1   1.5724279
142     70.9    63.8    0.5       1   1   0.7194403
144     96.2    57.5    0.8       1   1   2.3794853
145     76.1    79.4    9.9       1   1   1.0606354
147     41.6    70.7    0.8       0   1  -1.2030624
148     70.2    82.0   10.4       1   1   0.6735102
```

The vector of standardized achievement scores from this newly ordered data frame can then be used to test the `mean.diff()` function. Command Snippet 7.17 shows the results of executing the `mean.diff()` function on the standardized achievement scores from `ordered.latino`. This result is consistent with the difference in the observed conditional means.

Command Snippet 7.17: Executing the function `mean.diff()` on the vector of standardized achievement scores from the ordered data frame.

```
> mean.diff(ordered.latino$z.achieve)
[1] 0.3885426
```

7.13 FURTHER READING

A historical outline of bootstrap methods can be found in Hall (2003). Earlier work on resampling methods can be found in Quenouille (1949, 1956) and Tukey (1958). A very readable account of the bootstrap methodology can be found in Boos (2003). Theoretical foundations for the bootstrap methods can be found in Efron (1979) and Efron (1982), both of which are quite mathematical. Efron and Tibshirani (1993) and Davison and Hinkley (1997) provide a more thorough treatise of bootstrap methods with the latter providing the framework for their implementation through the **boot** package in Canty (2002), which includes examples. The treatment of missing data using bootstrap methods are broached in Efron (1994). For additional reading concerning Gosset and his work on the *t*-distribution, both Senn (2008) and Tankard (1984) provide good starting points, and the paper that started it all should not be overlooked (Student, 1908).

PROBLEMS

7.1   The National Center for Education Statistics (NCES) is mandated to collect and disseminate statistics and other data related to education in the United States. To that end, NCES began collecting data on specific areas of interest including educational, vocational, and personal development of school-aged children, following them from primary and secondary school into adulthood. The first of these studies was the National Longitudinal Study of the High School Class of 1972 (NLS-72). High School and Beyond[7] (HS&B) was designed to build upon NLS-72 by studying high school seniors, using many of the same survey items as the 1972 study. A sample of $N = 200$ students from the 1980 senior class cohort from HS&B were obtained and are located in the *HSB.csv* data set.

Use a nonparametric bootstrap to test if there is a difference in the variances of science scores, between public and private school students. That is, test the null hypothesis, $H_0 : \sigma^2_{\text{Public}} = \sigma^2_{\text{Private}}$. Write up the results from the analysis as if you were writing the results section for a manuscript for publication in a journal in your substantive area.

---

[7]For more information about HS&B including access to data sets go to http://www.nber.org/~kling/surveys/HSB.htm.

7.2   A pivot statistic for examining the difference in variability between two populations is the F-statistic. The F-statistic can be computed as

$$F = \frac{\sigma_1^2}{\sigma_2^2},$$

where σ_1^1 and σ_1^2 are the variances for group 1 and group 2, respectively. Ratios near one would indicate no difference in variances in the two populations. Use a nonparametric bootstrap test to again test the hypothesis of no difference in the variance of science scores between public and private school students, but this time use the F-statistic. Write up the results from both analyses as if you were writing a manuscript for a publication in a journal in your substantive area.

7.3   Compare and contrast the results from the two analyses.

7.4   The "ideal" number of bootstrap replicates to have no simulation error is inf. *How large should one make R, the number of bootstrap replications in practice?* You will investigate this question empirically. Rerun the nonparametric bootstrap for the F-statistic using six different values for the number of bootstrap replicates, namely $R = 25, 50, 100, 250, 500,$ and 1000. Compute the p-value for the observed F-statistic using each of the R values. Compare these values and give some recommendation as to the number of bootstrap replicates needed.

CHAPTER 8

PHILOSOPHICAL CONSIDERATIONS

For the past 40 years there has been confusion on the difference between tests of significance and tests of hypotheses to the point where data interpretation is presented as an accept–reject process.

—O. Kempthorne & T. E. Doerfler (1969)

Chapters 6 and 7 introduced two methods—the randomization test and the bootstrap test—that can be used by researchers to examine questions of interest by providing a degree of statistical evidence based on observed data for or against a particular assumed hypothesis (i.e., the null hypothesis). One decision that researchers have to make in the process of analyzing their data is which of these methods should be used or should an entirely different method be used. This is just one of many decisions that researchers face in the process of data analysis.

In this chapter, a glimpse is offered at a few of the decisions that need to be made along the way regarding the choice of an analysis method and of a philosophical testing framework. It must be acknowledged that there are entire books written about each of these issues and in some cases entire books written about subtopics of these issues.

Comparing Groups: Randomization and Bootstrap Methods Using R **173**
First Edition. By Andrew S. Zieffler, Jeffrey R. Harring, & Jeffrey D. Long
Copyright © 2011 John Wiley & Sons, Inc.

This chapter offers a limited survey of the relevant issues and readers are encouraged to pursue secondary sources to more fully explore topics of interest. This chapter is different from the others in the monograph as it offers no analyses nor computing work. As such, no end of chapter problems are offered. In addition, there is no *Further Reading* section in this chapter. Rather, references are interspersed throughout the chapter and all offer more in-depth treatment of specific topics.

8.1   THE RANDOMIZATION TEST VS. THE BOOTSTRAP TEST

The randomization/permutation test and the bootstrap test were introduced in the previous two chapters as methods to test for group differences. Which method should be used? From a statistical theory point of view, the difference between the two methods is that the randomization method is conditioned on the marginal distribution under the null hypothesis. This means each permutation of the data will have the same marginal distribution. The bootstrap method allows the marginal distribution to vary, meaning the marginal distribution changes with each replicate data set. If repeated samples were drawn from a larger population, variation would be expected in the marginal distribution, even under the null hypothesis of no difference. This variation in the marginal distribution is not expected; however, there is only one sample from which groups are being randomly assigned so long as the null hypothesis is true. Thus the choice of method comes down to whether one should condition on the marginal distribution or not.

The choice of whether to condition on the marginal distribution has been debated; see Upton (1982), Yates (1984), Rice (1988), D'Agostino (1998), and Little (1989). The key question addressed in all of these articles is, how is the variation in the chosen test statistic (i.e., the standard error) being computed? As Westfall and Young (1993) point out, this question is directly tied to the question "what is the scope of inference?"

In Chapter 6, four research scenarios were presented in which the method of obtaining a sample of data and the method used to assign subjects to treatments was either random or not. It was pointed out that the inferences a researcher could make were different in each of these situations. Table 8.1 re-presents these four scenarios but includes the valid scope of inference for each.

As an illustration, consider the After-School Program data, which had random assignment, but not random sampling, depicted in row 2 of Table 8.1. If evidence had been found against the null hypothesis, it would have been valid to suggest that the after-school program was better than the "treatment as usual" in terms of lowering delinquency rates. It would not have been valid to say that this was the case for different populations of students.

For the Latino achievement data, which included random selection but not random assignment, the inference that Mexican and non-Mexican immigrants differ in their levels of achievement can be extended to the larger population of Latino immigrants in the Los Angeles area (Table 8.1 row 1). However, that difference cannot be attributed in a causal manner to the difference in areas from which the persons emigrated.

Table 8.1: Four Potential Scenarios Researcher Could Face When Making Inferences[a]

Type of Research	RS[b]	RA[c]	Scope of Inference
1. Generalizable	✓		Population
2. Randomized Experimental		✓	Sample
3. Generalizable, Randomized Experimental	✓	✓	Population
4. Non-Generalizable, Non-Experimental			Sample

[a] The valid scope of inference is presented for each scenario.

[b] RS=Random sample

[c] RA=Random assignment

In light of Table 8.1, the choice of method for the data analysis—the randomization test or the bootstrap test—seems obvious in these two research scenarios. In the other two situations of Table 8.1, where both the selection and assignment of subjects are random (row 3) or both the selection and assignment are not random (row 4), the choice is not as clear. Consider the case where subjects are randomly sampled from a population and randomly assigned to treatments. The choice of analysis method rests solely on the scope of inferences the researcher wants to make. If inferences to the larger population are to be made, then the bootstrap method should be used, as it is consistent with the idea of sample variation due to random sampling. In general, there is more variation in a test statistic due to random sampling than there is due to random assignment. That is, the standard error is larger under the bootstrap. Thus, the price a researcher pays to be able to make broader inferences is that all things being equal, the bootstrap method will generally produce a higher p-value than the randomization method.

In many observational studies, the observed data can be treated as one of the many potential random samples that could have been drawn from the population(s). In these cases, the variation in the test statistic needs to be consistent with the variation produced through the sampling process. Because of this, the bootstrap test has been recommended for use with observational data. Using bootstrap resampling with observational data, however, is not a magic bullet that changes the scope of the inferences. Care needs to be taken when the results are being interpreted so that the researcher does not overgeneralize.

8.2 PHILOSOPHICAL FRAMEWORKS OF CLASSICAL INFERENCE

The method of statistical testing that has made its way into most introductory statistics textbooks is a hybrid of two very different philosophical frameworks. The approach that is typically presented is an amalgam of the ideas posited by R. A. Fisher (1890–1929) on one hand and Jerzy Neyman (1894–1981) and Egon Pearson (1895–1980) on the other. It is ironic that this combination has taken place considering that the philosophies are in stark contrast to one another. What has led to considerable confusion is that p-values, confidence intervals, and the like are used in both frameworks. What is different is the interpretation of these quantities.

8.2.1   Fisher's Significance Testing

Fisher's experience as a scientist and his views on inductive inference led him to take an objective and practical approach to statistical testing. His inductive inference is the philosophical framework that allows a researcher "to argue from consequences to causes, from observations to hypotheses" (Fisher 1966, p. 3). Fisher referred to his testing framework as *inductive inference*. It was in contrast to the earlier work of Thomas Bayes (1702–1761) and Pierre-Simon Laplace (1749–1827), who developed the idea of *inverse probability*.[1] The object with inverse probability was to provide the probability of a hypothesis (H) given the data

$$\mathbb{P}\left(H \mid \text{Data}\right). \tag{8.1}$$

Fisher rejected this idea and instead embraced the method of *direct probability*, which is based on the probability of the data given a particular hypothesis,

$$\mathbb{P}\left(\text{Data} \mid H\right). \tag{8.2}$$

Central to Fisher's method of inductive inference was the specification of what he referred to as the *null hypothesis*, H_0. Using a known distribution of a test statistic, T, under the assumption detailed in the null hypothesis, Fisher then assigned a probability to the same statistic computed in the observed data by determining the density in the test distribution that exceeded the observed value of T. This probability value (i.e., achieved significance level) could then be used as a strength of evidence argument against the null hypothesis.

The null hypothesis could be "rejected" if the strength of evidence met a certain degree of significance. Fisher set this degree at 5% writing, "It is usual and convenient for experimenters to take 5 percent, as a standard level of significance, in the sense that they are prepared to ignore all results which fail to reach this standard" (Fisher, 1966, p. 13). If the achieved significance level fell below this mark, the null hypothesis would be rejected, or in his words, "Every experiment may be said to exist only in order to give the facts a chance of disproving the null hypothesis" (Fisher, 1966, p. 16). But in other places he suggested 1%, or advocated even greater flexibility, depending on the situation.

---

[1] Inverse probability gave rise to a whole different philosophical framework and set of methods for inference referred to as *Bayesian* methods. Chatterjee (2003) provides a detailed summarization of the Bayesian approach to statistical induction within the historical evolution of statistical thought.

8.2.2 Neyman–Pearson Hypothesis Testing

To Neyman and Pearson, one question that Fisher had not answered sufficiently was why a particular test statistic, say T_1, should be used rather than another say T_2. Neyman and Pearson sought to address this issue in their work (e.g., Neyman & Pearson, 1928a, 1928b, 1933). Because Fisher's work was a touchstone, Neyman and Pearson initially viewed their method as building and expanding on Fisher's approach to statistical testing. Their solution for choosing a test statistic involved the statement of both the null hypothesis, as well as a class of alternative hypotheses. In contrast, Fisher conceived only of a null hypothesis.

The Neyman–Pearson approach was based on *inductive behavior* rather than inductive inference. Inductive behavior focued on rules for deciding between the two complementary hypotheses. The decision did not explicitly involve an evaluation of evidence based on sample data, as emphasized by the single-hypothesis framework of Fisher. In explaining the concept of inductive behavior, Neyman (1950, p. 1, 259–260) writes, "[inductive behavior] may be used to denote the adjustment of our behavior to limited amounts of information. The adjustment is partly conscious and partly subconscious. The conscious part is based on certain rules (if I see this happening, then I do that) which we call rules of inductive behavior." Explaining further that "to accept a hypothesis H means only to decide to take action A rather than action B. This does not mean that we necessarily believe that the hypothesis H is true ... [and to reject H] ... means only that the rule prescribes action B and does not imply that we believe that H is false."

Once a decision to take an action has been made, Neyman and Pearson acknowledged that an error could have occurred. Thus, it was important to quantify the probabilities associated with making two types of errors regarding the null hypothesis, namely a false rejection ("error of the first type") and false acceptance ("error of the second type"). This specification of error probabilities is what led them to a solution as to which test statistic should be used. The optimal test statistic is one that minimizes the error in falsely accepting the null hypothesis (Type II) while at the same time keeping the probability of falsely rejecting the null hypothesis (Type I) at a particular set bound.

Once the bound for the probability of falsely rejecting the null hypothesis, also called the significance level, has been determined, then the probability of falsely accepting the null hypothesis can be computed for each of the alternative hypotheses, as well as the power for each test. Power is the probability of correctly rejecting the null hypothesis as a function of the alternative hypothesis being tested. This probability can be used in "assessing the chance of detecting an effect (i.e., a departure from H) when it exists, determining the sample size required to raise this chance to an acceptable level, and providing a criterion on which to base the choice of an appropriate test" (Lehmann, 1993, p. 1244).

8.2.3 *p*-Values

One issue at the heart of the difference between Fisher and Neyman–Pearson is the interpretation of the *p*-value. Fisher interpreted the *p*-value as a sample-based measure of evidence against the null hypothesis. Neyman–Pearson regard it as a device for making a decision for one of the two hypotheses (and against the other). The point is that in Fisher's system, the size of the *p*-value matters, as it conveys effect size information regarding the discrediting of the null hypothesis. In the Neyman–Pearson system, the exact *p*-value is inconsequential, only that it is above or below a particular threshold (Hubbard & Bayarri, 2003).

This inconsistency in interpretation also manifests itself in the reporting of research results. Style guides, such as the *Publication Manual* of the APA, appear to suggest reporting practices that would satisfy neither Fisher nor Neyman and Pearson. For example, asterisks are suggested in tables to denote statistical significance (surpassing a threshold), but there is also a recommendation that the exact *p*-value be reported. This marriage of incompatible philosophies and methodologies has led to confusion and misinterpretation of the results by researchers and consumers of research. These problems have been widely documented and criticized (e.g., Cohen, 1990, 1994; Kupfersmid, 1988; Rosenthal, 1991; Rosnow & Rosenthal, 1989; Shaver, 1985; Sohn, 2000; Thompson, 1994, 1997).

For applied researchers, the Fisherian philosophy of testing and *p*-value interpretation appears to have advantages. The authors of this monograph believe, as Fisher did, that the scientific process is primarily about evaluating evidence, not just an exercise in decision making. Under Fisher's framework, the *p*-value assists the learning process by serving as a continuous measure of evidence in the larger "programmatic" sense. This is more sensible than trying to draw conclusions from the results of a single study, the absolutism of declaring a treatment "significant" or "not significant'," based on what is often an arbitrary threshold. When Fisher used 1 in 20 as a benchmark, he realized that if a study reported a *p*-value of 0.06 and the treatment in question was truly useful, another researcher might replicate the results and show it to be useful. Thus, rather than tossing out the treatment altogether under the guise of "nonsignificant," "marginal *p*-values" suggest that the effect of the treatment could be studied further using a better research design.

Adopting a Fisherian stance, the following recommendations are offered for reporting and interpreting the results from statistical tests.

- **Treat the *p*-value as a sample-based measure of evidence against the null hypothesis.** Report the numerical *p*-value rather than the more ambiguous inequality statements, such as $p \leq .05$. Try to refrain from including asterisks next to results. Avoid phrases such as, "the result is statistically significant," which implies that a definitive decision was made. An alternative is to use the term "statistically reliable" when the *p*-value is sufficiently small (see below).

- **Offer qualitative descriptions along with the *p*-value to emphasize that the results are part of the larger corpus of the scientific process.**. Efron and Gous (2001) offer descriptions of effect size for various *p*-values based on

Fisher's writings. Table 8.2 presents these descriptions. Based on the table, a p-value of 0.06 is considered "weak" evidence, which is very different than "no" evidence.

Table 8.2: Fisher's Descriptions for Scale of Evidence Against the Null Hypothesis[a]

Strength of Evidence	p-Value
Borderline or weak	0.100
Moderate	0.050
Substantial	0.025
Strong	0.010
Very strong	0.005
Overwhelming	0.001

[a] Based on a table presented in Efron and Gous (2001).

Based on these recommendations, the write-up for the Latino achievement analysis might be revised to read as follows:

Sample Write-Up

A nonparametric bootstrap test was used to compare the educational achievement between U.S. immigrants from Mexico and U.S. immigrants from other Latino countries. The data were studentized prior to the analysis. A nonparametric bootstrap test was used to determine whether there was a **statistically reliable** difference in the mean achievement scores between Mexican and non-Mexican immigrants. A Monte Carlo p-value was computed by drawing 4999 bootstrap replicates of the data. Using a correction suggested by Davison and Hinkley (1997), a p-value of 0.048 was computed. **This is moderate evidence against the null hypothesis of no population differences (Efron & Gous, 2001), and may suggest that Mexican and non-Mexican immigrants really do differ in their educational achievement.**

CHAPTER 9

BOOTSTRAP INTERVALS AND EFFECT SIZES

One of the great questions in statistical inference is: How big is it?

—J. Simon (1997)

In Chapters 6 and 7, the use of the randomization/permutation test and the bootstrap test were introduced as methods that a researcher can use to examine group differences. But, as Kirk (2001) points out, the group differences question often expands into three questions: (a) Is an observed effect real or should it be attributed to chance? (b) If the effect is real, how large is it? and (c) Is the effect large enough to be useful? Using the inferential methods introduced thus far, only the first of Kirk's three questions can be answered. As noted by Pearson (1901) early in the history of statistics, the results from inferential tests provide only the answer to the first question and therefore must be supplemented.

Recently, numerous researchers have argued for the use of *effect sizes* to complement or even replace tests of significance (Cohen, 1990, 1994; Kirk, 1995, 1996; Thompson, 1996, 2007). Effect size is a term used to describe a family of indices that characterize *the extent to which sample results diverge from the expectations specified in the null hypothesis.* These measures help researchers focus on the meaningfulness of significance testing results by providing answers to Kirk's second two questions

Comparing Groups: Randomization and Bootstrap Methods Using R
First Edition. By Andrew S. Zieffler, Jeffrey R. Harring, & Jeffrey D. Long

and also provide a method by which to compare the results between different studies. Effect sizes have been fully embraced by the APA. Its *Publication Manual* (American Psychological Association, 2009, p. 25) states, "For the reader to fully understand the importance of your findings, it is almost always necessary to include some index of effect size or strength of relationship in your Results section." It further states that the "failure to report effect sizes" may be found by editors to be one of the "defects in the design and reporting of results" (p. 5).

9.1 EDUCATIONAL ACHIEVEMENT AMONG LATINO IMMIGRANTS: EXAMPLE REVISITED

Consider, again, the data on the educational achievement among Latino immigrants—found in LatinoEd.csv. In Chapter 7, the question of whether there was a difference in the educational achievement of immigrants from Mexico and that of immigrants from other Latin American countries was examined. Moderate evidence against the null hypothesis of no population differences, $p = 0.045$, was found, using the nonparametric bootstrap test. This suggests that Mexican and non-Mexican immigrants likely differ in their educational achievement. Consider what is known and not known at this point.

- There is statistical evidence of differences in educational achievement between the population of Latino immigrants from Mexico and Latino immigrants from other countries.

- However, it is not known how the populations differ, in term of the direction of the difference.

- Nor is the magnitude of the difference known.

Using the sample evidence (e.g., sample means), more insight into how the two groups differ can be provided. For example, the mean educational achievement score for Latino immigrants from Mexico is 58.6 and that for non-Mexican immigrants is 64.5. The sample evidence suggests that non-Mexican immigrants have, on average, a higher level of educational achievement than Mexican immigrants. The magnitude of the difference is 5.9 achievement points. Using substantive knowledge, one could then make decisions about whether a difference this large is meaningful or not.

9.2 PLAUSIBLE MODELS TO REPRODUCE THE OBSERVED RESULT

The magnitude of the difference between the two populations is estimated using the sample mean difference. This is an estimate of the population effect,

$$\widehat{\text{Effect}} = \left| \bar{Y}_{\text{Mexican}} - \bar{Y}_{\text{non-Mexican}} \right|$$
$$= |64.5 - 58.6|$$
$$= 5.9.$$

The conclusion is that the estimated educational achievement for Latino immigrants from Mexico is, on average, 5.9 points lower than for Latino immigrants from other countries. This is a model for the true population mean difference in educational achievement between Latino immigrants from Mexico and Latino immigrants from other countries. This model can be expressed as

$$\text{Model 1}: \mu_{\text{non-Mexican}} - \mu_{\text{Mexican}} = 5.9, \quad \text{or}$$
$$: \mu_{\text{Difference}} = 5.9.$$

It is stressed that Model 1 is a hypothesis, as the true population difference cannot be determined based on sample data. The p-value produced had this model been tested, rather than the model having a population mean difference of zero, would not have provided evidence against the model. That is because the model having the parameter value of 5.9 would reproduce the observed mean difference with a high likelihood. Might there be other models that would also reproduce the observed mean difference of 5.9 with reasonably high likelihood?

Figure 9.1 shows the density plot of the bootstrap distribution of the mean difference for three other models (hypotheses). The first bootstrap distribution (dotted line) shows the distribution of mean differences based on the model in which the population mean difference is assumed to be 0. The second bootstrap distribution (dashed line) shows the distribution of mean differences based on the model in which the population mean difference is assumed to be 3. The third bootstrap distribution (dot-dashed line) shows the distribution of mean differences based on the model where the population mean difference is assumed to be 10. The point in the plot demarcates the observed mean difference in the data of $\bar{Y}_{\text{non-Mexican}} - \bar{Y}_{\text{Mexican}} = 5.9$. The second model, which has the parameter value of 3, has the highest likelihood of reproducing the sample mean difference. This is illustrated by the fact that the point in Figure 9.1 is in the neighborhood of the peak of the distribution. Compare this with the model of no difference and also the model having a mean difference of 10. These two models have a lower likelihood of reproducing the sample mean difference.

These plots illustrate the following.

- Different models (or hypotheses) for the population mean difference will generate different distributions of the mean differences.

- Some models will generate simulated data that reasonably reproduce the observed result. These models can be considered as plausible.

- Other models will not generate simulated data that seem to reasonably reproduce the observed result. These models should probably be considered as implausible.

9.2.1 Computing the Likelihood of Reproducing the Observed Result

Of the three models presented in Figure 9.1, it is evident that the model having a parameter value of 3 is a more plausible candidate for the model that generated the

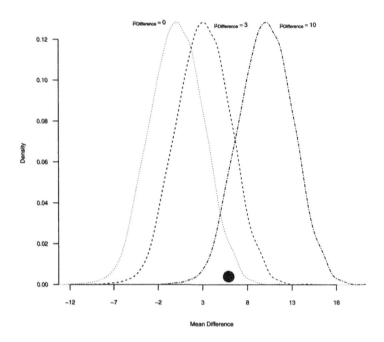

Figure 9.1: The bootstrap distributions of the standardized mean difference based on the model with parameter value 0 (dotted), parameter value 3 (dashed), and parameter value 10 (dot-dashed). The point in the plot demarcates the observed mean difference in the data of 5.9.

sample result of 5.9 than the other two models. Figure 9.1 clearly illustrates that there is substantial separation between the center of the distribution for the model of no population mean difference (dotted line) and the observed mean difference of 5.9. However, the separation is not always so dramatic. When the separation is not so clear, it is helpful to quantify the likelihood of a particular model in generating the observed data.

To quantify the likelihood of reproducing the observed data, a one-sided p-value can be computed for each potential model. The p-value is one-sided because the discrepancy between the observed result and the model's parameter value is being measured in a particular direction (in terms of the figure, either to the left or to the right). For example, in the model which has a parameter value of 3 (dashed line) in Figure 9.1, the parameter value is smaller than the observed mean difference of 5.9. Thus, the p-value for this model is quantifying the probability of obtaining a mean difference *as large or larger than* the observed mean difference of 5.9. Since the observed mean difference is in the right-hand side of the bootstrap distribution for this model, the *p*-value needs to measure the discrepancy from the parameter value in this direction. The model which has the parameter value of 10 (dot-dashed line) in Figure 9.1, on the other hand, has a parameter value that is larger than the observed mean difference of 5.9. Thus, the *p*-value for this model is quantifying the probability of obtaining a mean difference *as small or smaller than* the observed mean difference of 5.9.

To examine the likelihood of a potential model to reproduce the observed data, the parameter value under consideration is added to the mean difference computed in the function used in the nonparametric bootstrap. Command Snippet 9.1 shows the syntax for bootstrapping the mean difference under the model that the population mean difference is 3. (See Chapter 7 for details on the nonparametric bootstrap.)

Command Snippet 9.1: Syntax for bootstrapping the mean difference under the model that the population mean difference is 3. The bootstrap distribution is plotted along with the observed mean difference of 3. The one-sided *p*-value is also computed.

```
## Function to compute the mean difference under the model
## with the parameter value of 3
> pv.3 <- function(data, indices) {
  d <- data[indices,]
  mean(x = d$Achieve[1:34]) - mean(x = d$Achieve[35:150]) + 3
  }

## Carry out the nonparametric bootstrap
> model.boot <- boot(data = latino, statistic = pv.3, R = 4999)

## Compute the mean of the bootstrap distribution
> mean(model.boot$t)
[1] 3.068308

## Compute the SE for the bootstrap distribution
> sd(model.boot$t)
[1] 2.992834

## Count the bootstrapped mean differences at or below 5.9
> length(model.boot$t[model.boot$t >= 5.9])
[1] 875

## Compute the one-sided p-value
> (875 + 1)/(4999 + 1)
[1] 0.175
```

Figure 9.2 shows a plot of the bootstrap distribution. The vertical dashed line at three demarcates the model's parameter value. The plot is centered at the parameter value, as also evidenced by the mean of the bootstrapped mean differences in Command Snippet 9.1. The standard error of the distribution is identical to the standard error of the distribution that assumed no population mean difference (see Chapter

7). The one-sided p-value of 0.175 suggests that this model has some likelihood of reproducing the observed data. That is, a sample value of 5.9 is not a particularly rare event under the model (or hypothesis) that the population difference is 3.

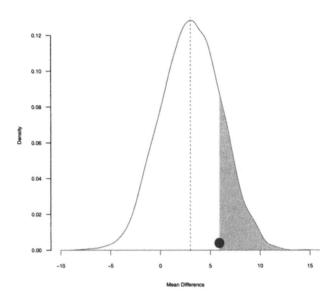

Figure 9.2: The bootstrap distribution of the mean difference based on the model with parameter value 3. A vertical dashed line is drawn at 3, the model's parameter value. The point in the plot demarcates the observed mean difference in the data of 5.9. The one-sided p-value, indicating the likelihood of this model to reproduce the observed data, is shaded.

Table 9.1 presents the parameters for several potential models for reproducing the observed data. The one-sided p-value for each model is also presented. The goal of such an analysis is to identify a range of plausible models that would reproduce the observed data. Lastly, the qualitative descriptors from Efron and Gous (2001) are listed to describe the strength of evidence against each model.

The models that have parameter values between 2 and 10 seem to plausibly reproduce the observed mean difference of 5.9. By this it is meant that the single sample results of 5.9 is consistent with parametric bootstrap models that have a mean of 2 to 10. The plausibility criterion here is that the p-value provides less than a moderate degree of evidence against the model ($p > 0.05$), though other levels of evidence could be used. It should also be mentioned that Table 9.1 is based on parameter values that are integer. Greater continuity in the p-values is achieved when fractions are considered for the parameter value (e.g., 1.1, 1.2, etc.).

Recall that the parameter value indicates the population mean difference in the achievement scores between the non-Mexican and Mexican immigrants. The range of parameter values that plausibly reproduce the observed result can then be considered candidates for the true population mean difference. The range of $[2, 10]$ is an *interval estimate*—often referred to as a confidence interval—for the population mean difference. It provides a range of plausible values for the sample estimate, accounting for the expected variation due to either random sampling or random assignment. Though not pursued here, greater precision is gained by considering fractional parameter values near the endpoints (e.g., 2.5 and 9.5, etc.).

It is also noted that the data produce the weakest degree of evidence (highest p-value) against the model having a parameter value of six. The model having the parameter value of the sample mean difference, 5.9, will produce the highest p-value. That is because this model has the highest likelihood of reproducing the observed mean difference. As the parameter value gets further away from the observed mean difference of 5.9, the models are less likely to plausibly reproduce the observed data.

Table 9.1: Parameter Value, p-Value, and Qualitative Description of Strength of Evidence Against the Model for 11 Potential Models

Model Parameter	p-Value	Strength of Evidence
0	0.026	Substantial
1	0.051	Moderate
2	0.099	Borderline
3	0.175	Weak
4	0.274	Negligible
5	0.397	Negligible
6	0.475	Negligible
7	0.347	Negligible
8	0.238	Negligible
9	0.149	Weak
10	0.086	Borderline
11	0.039	Substantial
12	0.017	Strong

[a] The p-value for each model is based on drawing 4999 bootstrap replicates.

9.3 BOOTSTRAPPING USING AN ALTERNATIVE MODEL

The search method for obtaining the range of plausible values just presented has two issues. First, it is unclear what range of values of the mean difference should be used.

Second, it can be very time consuming to test the large number of models required for greater accuracy. As mentioned, only integer parameter values were considered in Table 9.1 and a more accurate assessment would involve fractional values. A useful shortcut to finding the interval estimate is to bootstrap the mean difference under the assumption that there is a difference between the two groups. In effect, this approach is equivalent to the approach taken in Table 9.1, but using all possible parameter values in some interval, say 2 to 10.

Chapter 7 discussed the testing of the hypothesis of no average achievement difference using bootstrap replicates drawn under the auspices of the null hypothesis. In this scenario, the observations were pooled and then the bootstrap replicates were drawn from the pooled sample. Now the strategy is changed so that the bootstrap is conducted under the auspices of an alternative hypothesis. It is now assumed that the samples represent two different populations with two different mean values.

In these situations, a plausible alternative model is "automatically" selected based on the following nonparametric bootstrap method. Since it is believed that the samples represent two different populations with two different mean values, the replicate data sets are drawn from the two samples separately—the first replicate sample is bootstrapped only using the observations from the first sample, and the second replicate sample is bootstrapped only using the second observed sample. The groups are kept separate, meaning that randomization occurs only within a group—not among the groups—which yields an estimate of the distribution of mean differences under the assumption that the populations are different. Thus, the method can be thought of as an evaluation of a range of general null hypotheses. This method is convenient as it, in effect, produces results similar to those in Table 9.1, but uses continuous parameter values rather than just integer values.

Figure 9.3 represents the concept of the nonparametric bootstrap under an alternative model. Consider two samples of data, $\{x_1, x_2, x_3, \dots x_{n_1}\}$ and $\{y_1, y_2, y_3, \dots y_{n_2}\}$, each drawn from an infinitely large population where both probability distributions are unknown. Because these probability distributions are unknown, the empirical distributions based on the observed data can be used as representations of the two population distributions. Figure 9.4 shows the steps to bootstrap the distribution for a particular test statistic, $\hat{V}$.

Consider how a nonparametric bootstrap would be used to draw replicate data sets under the alternative model in order to estimate the model parameter values that would plausibly reproduce the observed mean difference of 5.9. Recall that in the observed data, there were 116 Mexican immigrants and 34 non-Mexican immigrants.

- First, resample with replacement 34 observations from the sample of only the non-Mexican immigrants achievement scores. Then, resample with replacement 116 observations from the sample of achievement scores for only the Mexican immigrants. This is the bootstrap replicate data set.

- Calculate the mean difference for the bootstrap replicate data set by subtracting the mean for the resampled Mexican achievement scores from the mean for the resampled non-Mexican achievement scores (or vice-versa).

- Repeat the process many times and record the mean difference each time. The resulting collection of means differences constitute the bootstrap distribution of the mean differences under the alternative hypothesis.

- Use the bootstrap distribution to determine a plausibility range of values for the population parameter.

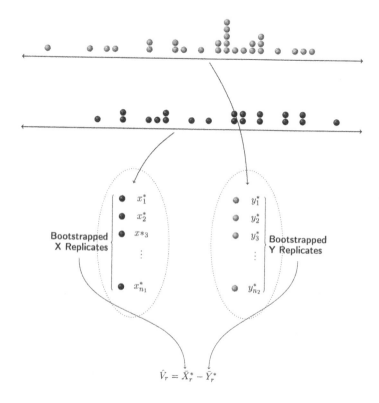

Figure 9.3: Nonparametric bootstrapping under an alternative model involves re-sampling from the observed sample with replacement within defined strata.

9.3.1   Using R to Bootstrap under the Alternative Model

It is worth reiterating that the difference between the bootstrap carried out under the null hypothesis and the bootstrap carried out under the alternative model is that the latter does not combine the data of the groups prior to resampling. The groups are left intact (separate) for the resampling. To bootstrap under the general alternative model in R, the **boot** package is again used. Command Snippet 9.2 includes a function called mean.diff.alt() that when applied to the Latino data will compute the mean

---

Nonparametric Bootstrapping under Alternative Model

- Randomly resample n_1 observations from the first observed sample *with replacement* $\{x_1^*, x_1^*, x_2^*, x_3^*, \ldots, x_{n_1}^*\}$, and resample n_2 observations from the second observed sample *with replacement* $\{y_1^*, y_1^*, y_2^*, y_3^*, \ldots, y_{n_2}^*\}$.

- Compute $\hat{V}_r$—the test statistic of interest—using the bootstrapped observations $\{x_1^*, x_1^*, x_2^*, x_3^*, \ldots, x_{n_1}^*\}$ and $\{y_1^*, y_1^*, y_2^*, y_3^*, \ldots, y_{n_2}^*\}$ (e.g., $\hat{V}_r$ might be the mean difference).

- Repeat the first two steps many times, say R times, each time recording the value of the statistic $\hat{V}_r$.

- The distribution of $\hat{V}_1, \hat{V}_2, \hat{V}_3, \ldots, \hat{V}_R$ can be used as an estimate of the sampling distribution of $\hat{V}$ under the alternative model.

Figure 9.4: Steps for using the sample data to bootstrap the distribution for a particular test statistic, $\hat{V}$, under an alternative model.

difference between the achievement scores for the resampled Mexican immigrants and the resampled achievement scores for the resampled non-Mexican immigrants. Because the bootstrapping needs to be carried out within groups, the manner in which the groups were formed in Chapter 7—using the first 116 observations sampled and the last 34 observations—can no longer be used. Instead, the groups are indexed using the Mex variable.

Command Snippet 9.2: A function to compute the mean difference between the achievement scores for the resampled Mexican immigrants and the resampled achievement scores for the resampled non-Mexican immigrants. The first line indicates that all rows of the data frame are to be used.

```
> mean.diff.alt <- function(data, indices) {
    d <- data[indices, ]
    mean(d$Achieve[d$Mex == 0]) - mean(d$Achieve[d$Mex == 1])
    }
```

Since the groups are being indexed via the Mex variable, the function can be tested on the latino data frame without having to reorder the data as was shown in Section 7.12. Command Snippet 9.3 shows the use of the mean.diff.alt() function on the latino data frame. The function correctly returns the observed mean difference of 5.9 (within rounding).

Command Snippet 9.3: The use of `mean.diff.alt()` on the `latino` data frame to test that it is computing the mean difference correctly.

```
> mean.diff.alt(latino)
[1] 5.921602
```

After executing this function in R, it is used as an argument in the `boot()` function as `statistic=mean.diff.alt`. The optional argument `strata=` is also included so that the bootstrapping is performed *within each strata* or group. Command Snippet 9.4 shows the syntax for using the `boot()` function to bootstrap the mean difference in achievement scores for 4999 replicates. The results are assigned to an object called `altmodel.boot`.

Command Snippet 9.4: Bootstrapping the mean difference in achievement scores between Mexican and non-Mexican immigrants for 4999 replicates. The bootstrapping is carried out within strata rather than by pooling as in the last chapter. These results are also plotted and displayed.

```
## Bootstrap under the alternative model
> altmodel.boot <- boot(data = latino, statistic =
    mean.diff.alt, R = 4999, strata = latino$Mex)

## Plot the bootstrap distribution of mean differences
> plot(density(altmodel.boot$t))

## Mean of the bootstrapped mean differences
> mean(altmodel.boot$t)
[1] 5.946968

## Standard error of the bootstrapped mean differences
> sd(altmodel.boot$t)
[1] 2.631824
```

Figure 9.5 shows the plotted results of the bootstrapped mean difference under the alternative model. Unlike the plot under the null model, this distribution is centered at 5.9, the observed mean difference, rather than at 0. The estimated standard error is smaller than the estimated standard error based on the null model. This is because the resampling was carried out within each group separately, it was *stratified*. Stratified sampling reduces the standard error when the variation of scores within each group is less than the variation in the pooled scores (Lohr, 2010).

There is still one issue that is left unaddressed. The height of the bootstrap distribution in Figure 9.5 is an indication of the plausibility of sample values. As the values spread out from the center, they become increasingly implausible. It is usually desirable to establish limits for plausibility by determining endpoints for a certain area under the bootstrap curve. This topic is now addressed.

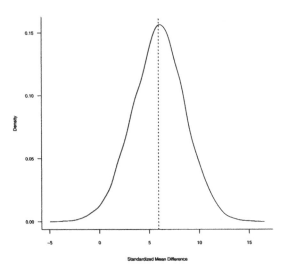

Figure 9.5: Bootstrap distribution of the mean difference in educational achievement scores between non-Mexican and Mexican immigrants for 4999 replicates. The resampling was carried out under the alternative model, thus centering the distribution around the observed mean difference of 5.9 (dashed line).

9.3.2 Using the Bootstrap Distribution to Compute the Interval Limits

The computation of the endpoints, or *limits*, for the interval estimate is fundamentally based on the *sampling distribution* of the statistic of interest (e.g., the distribution of *all* possible standardized mean differences). The sampling distribution is typically used to compute the standard error of the statistic of interest, which is then used to compute the interval limits.

9.3.3 Historical Interlude: Student's Approximation for the Interval Estimate

Historically, researchers in the educational and behavioral sciences have used Student's t-distribution as an approximation to the sampling distribution for many statistics (e.g., mean, mean difference, etc.). This distribution is used because under certain assumptions—repeated, independent sampling from normally distributed populations— the probability density of the t-distribution is known and can be used to compute the standard error of the estimate.

Consider the analytic formula for t,

$$t = \frac{\hat{V} - E(V)}{\text{SE}_{\hat{V}}},$$ (9.1)

where $\hat{V}$ is an observed statistic that is being used to estimate a parameter $E(V)$, and $\text{SE}_{\hat{V}}$ is the estimated standard error of the sampling distribution of $\hat{V}$. Using Equation 9.1, it can be shown that the interval limits are computed as

$$\hat{V} \pm B \times \text{SE}_{\hat{V}},$$ (9.2)

where B is the quantile from the t-distribution which demarcates a threshold for the strength of evidence. This quantile is chosen to differentiate between a model that plausibly reproduces the observed result, and one that does not. In the educational and behavioral sciences, the threshold value is often set near the value of two. This is because in a t-distribution, the quantile of two produces the approximate rule of $p \leq 0.025$ (substantial evidence against the model) for eliminating a model as a candidate.

9.3.4 Studentized Bootstrap Interval

Consistent with the overarching philosophy of this monograph, bootstrap intervals are emphasized over the completely analytic approach of Equation 9.1. One method of obtaining bootstrap intervals is the studentized bootstrap method. The *studentized bootstrap method* has the same form as Equation 9.2, but the estimate of $\text{SE}_{\hat{V}}$ is obtained from the bootstrap results. From the results printed in Command Snippet 9.4, the estimate for the SE for the mean difference is 2.6. To construct the interval, the observed mean difference of 5.9, $B = 2$ and the bootstrap SE of 2.6 are used. Substituting these into Equation 9.2 yields

$$5.9 \pm 2 \times 2.6$$
$$5.9 \pm 5.2.$$

This gives an interval estimate of $[0.7, 11.1]$. Having computed the interval, it is important to discuss the meaning of the values in brackets. Therefore, issues of interpretation are now discussed.

9.4 INTERPRETATION OF THE INTERVAL ESTIMATE

There are different interpretations of the above interval depending upon the philosophical framework that is followed, as pointed out in Chapter 8. Under the Neyman–Pearson paradigm, the focus is on the relative frequency of the method to produce an interval that includes the parameter based on hypothetical repetitions of the process that produced the data (i.e., under repeated sampling from the population). This type of interpretation focuses on the *process* of constructing the interval under repeated sampling, rather than on the values of the single interval obtained in practice. For this reason, the Neyman–Pearson interpretation does not appear to provide statistical evidence for examining the plausibility of population parameters (Berger, 2003). In this sense, the Neyman–Pearson approach may not be particularly useful for applied research.

The Fisherian interpretation of the interval is evidence-based. The interval identifies parameter values for the population mean difference that are plausible candidates to reproduce the observed mean difference of 5.9. In other words, the interval gives the range of parameter values that are plausible candidates for the true mean difference. The interpretation of the interval is as a list of likely values for the population mean difference. For the example, the likely values for the mean difference between Mexican and non-Mexican Latino immigrants ranges from 0.7 to 11.1.

9.5   ADJUSTED BOOTSTRAP INTERVALS

In some situations, the interval produced using the studentized method can be biased or inaccurate. Many statisticians have proposed adjustments to the interval to improve particular properties (see Carpenter & Bithell, 2000; Good, 2005a, for a review of these methods). For example, Efron (1981a) proposed methods for adjusting the interval limits based on bias correction adjustments. Many other methods for improving the interval limits using bootstrap methods have been proposed. The section on Further Reading offers many places to start learning about these methods.

One method that tends to produce good interval limits in practice is the bootstrap *bias-corrected-and-accelerated* (BC_a) method. The calculation of the limits based on this method is beyond the scope of this monograph, but both Efron (1987) and DiCicco and Tibshirani (1987) provide more details for the interested reader.

The BC_a method of adjustment is implemented in the boot.ci() function from the **boot** package. This function takes the bootstrap object as a required argument. The optional argument type="bca" is used to specify that the bias-corrected-and-accelerated adjustment should be computed. Command Snippet 9.5 shows the syntax for finding the BC_a interval for the mean difference in educational achievement between non-Mexican and Mexican Latino immigrants.

The interval limits based on the BC_a method of adjustment are slightly different than those based on the studentized interval estimate. This method has resulted in a smaller range of values than Equation 9.2.

9.6   STANDARDIZED EFFECT SIZE: QUANTIFYING THE GROUP DIFFERENCES IN A COMMON METRIC

When the raw or original metric is being used to compute the effect, Kirk's question about the magnitude of the effect can be difficult to answer. For example, is the mean difference of 5.9 educational achievement points a small difference? A moderate difference? A large difference? Sometimes reporting raw estimates of effect can make sense from the standpoint that other researchers can relate to the metric being used. For example, many people can relate to a 7% difference in graduation rates because they understand the magnitude of a percent. However, other times the raw effect cannot be readily interpreted.

Command Snippet 9.5: Obtaining the limits for the nonparametric bootstrap BC_a confidence interval for the mean difference in achievement scores for 4999 replicates.

```
> boot.ci(boot.out = altmodel.boot, type = "bca")
BOOTSTRAP CONFIDENCE INTERVAL CALCULATIONS
Based on 4999 bootstrap replicates

CALL :
boot.ci(boot.out = altmodel.boot, type = "bca")

Intervals :
Level       BCa
95%    ( 0.553, 10.945 )
Calculations and Intervals on Original Scale
```

To alleviate the problems in interpretations of magnitude, researchers tend to report *standardized effect sizes* rather than raw effect sizes when the metric is not universally known. In considering differences between groups, there are two common standardized effect sizes that researchers use: distance measures and variance accounted for measures. While both of these measures characterize the extent to which sample results diverge from the expectations specified in the null hypothesis, they do so in a different manner. In this section, distance measures of effect are introduced. In Chapter 11, variance accounted for measures of effect size are discussed.

9.6.1 Effect Size as Distance—Cohen's δ

Jacob Cohen (1962) introduced the first commonly recognized effect size in an effort to call attention to statistical power in the behavioral sciences. Later, in his landmark 1965 book, he named this standardized effect size δ (pronounced "delta"). It has henceforth been referred to as Cohen's δ. The calculation of this standardized effect size is

$$\delta = \frac{|\mu_1 - \mu_2|}{\sigma} \tag{9.3}$$

where μ_1 and μ_2 are the two population means and σ is the standard deviation of the population. It is assumed the populations have different means (μ_1 and μ_2) but a common standard deviation, σ.

An estimate of this effect size is obtained by substituting the sample estimates for the mean and standard deviation into Equation 9.3. To identify this as an estimate, it is renamed d.

$$d = \frac{|\bar{Y}_1 - \bar{Y}_2|}{s} \tag{9.4}$$

Cohen pointed out that since the variances—and therefore the standard deviations—for the two populations are assumed to be equivalent, that the standard deviation estimate from either group could be used. However, in practice it is common to use the average of the two sample estimates, if the groups have equal sample sizes, or the pooled estimate,[1] if the groups do not have equal sample sizes. All that Equation 9.3 is doing is expressing the estimated effect between the population means of the two groups in terms of the size of the standard deviation—which is equivalent to computing the standardized mean difference.

To compute Cohen's d using R, the smd() function found in the **MBESS** library is used. The function takes six arguments, Mean.1=, Mean.2=, s.1=, s.2=, n.1=, and n.2=. These arguments take the values of the mean, standard deviation, and sample size for the two samples, respectively. The estimate of the standardized effect size is computed in Command Snippet 9.6.

Command Snippet 9.6: Computing the standardized effect size for the Latino data.

```
## Load MBESS package
> library(package = MBESS)

## Compute Cohen's d for latino data
> smd(
    Mean.1 = mean(latino$Achieve[latino$Mex == 0]),
    Mean.2 = mean(latino$Achieve[latino$Mex == 1]),
    s.1 = sd(latino$Achieve[latino$Mex == 0]),
    s.2 = sd(latino$Achieve[latino$Mex == 1]),
    n.1 = length(latino$Achieve[latino$Mex == 0]),
    n.2 = length(latino$Achieve[latino$Mex == 1])
    )
[1] 0.3924999
```

Note that the estimated value for Cohen's d is the same as the mean difference based on the standardized achievement scores that was computed in Chapter 7. The reason is that the standard deviation of the standardized achievement scores is equal to 1, and thus the value of Cohen's d is simply the mean difference. Cohen (1965) provided some guidelines for further interpretation of the magnitude of the computed standardized effect size. These are

$$d = .2 \quad \text{Small Effect Size,}$$
$$d = .5 \quad \text{Medium Effect Size, and}$$
$$d = .8 \quad \text{Large Effect Size.}$$

While it is common to provide an indication of the magnitude along with the computed effect size when reporting results, like all other guidelines, these should not be used without careful thought. In fact, Cohen (1965, p. 23) himself specified

---

[1]The pooled estimate is a weighted average based on the two groups' sample sizes.

that these conventions only made sense in the behavioral sciences, and even then cautioned researchers writing,

> The terms "small," "medium," and "large" are relative, not only to each other, but to the area of behavioral science or even more particularly to specific content and research method being employed in any given investigation.

9.6.2 Robust Distance Measure of Effect

Researchers for years have used the sample means and standard deviations in place of the parameters when estimating Cohen's δ. These estimates, however, are not robust, and therefore small changes in the distribution—such as a single extreme value—can greatly impact the estimates. Because of this issue, recent literature has suggested that in many situations using the sample means and standard deviation results in a systematically poor estimate of the true effect size (e.g., Staudte & Sheather, 1990; Wilcox, 2005). Algina, Keselman, and Penfield (2005) have proposed a solution to this problem by using robust estimates of the population means and standard deviation. Their estimate, called d_R, is computed as

$$d_R = .642 \left(\frac{|\bar{Y}_{t2} - \bar{Y}_{t1}|}{S_W} \right) \tag{9.5}$$

where $\bar{Y}_{t1}$ and $\bar{Y}_{t2}$ are the 20% trimmed means for the two groups. S_W is the square root of the pooled 20% Winsorized variance. The value, 0.642, is a scaling constant used to ensure that d_R and d are equal when the distributions for the two groups are normally distributed with equal variances. To compute d_R, the smd() function is again used. The robust estimate of the standardized effect size is computed in Command Snippet 9.7. The robust estimate is slightly smaller than Cohen's d.

Command Snippet 9.7: Computing the robust effect size for the Latino data.

```
## Load WRS library to use the winvar() function
> library(WRS)

## Compute robust effect size for latino data
> 0.642*smd(
      Mean.1 = mean(latino$Achieve[latino$Mex == 0], tr = 0.2),
      Mean.2 = mean(latino$Achieve[latino$Mex == 1], tr = 0.2),
      s.1 = sqrt(winvar(latino$Achieve[latino$Mex == 0], tr =
          0.2)),
      s.2 = sqrt(winvar(latino$Achieve[latino$Mex == 1], tr =
          0.2)),
      n.1 = length(latino$Achieve[latino$Mex == 0]),
      n.2 = length(latino$Achieve[latino$Mex == 1])
      )
[1] 0.3822657
```

Similar to an interval estimate for a sample statistic, an interval estimate for the effect size can be computed. The boot() and boot.ci() functions can be used to find an interval estimate for this robust measure of effect. The use of these functions to compute the limits for the nonparametric bootstrap BC_a interval is presented in Command Snippet 9.8. In this case, the estimated robust group difference is proposed to be as small as 0.0089 and as large as 0.7230.

Command Snippet 9.8: Function and nonparametric bootstrap for the robust standardized effect size for the mean difference in achievement scores based on 4999 replicates.

```
## Function for robust standardized effect
> robust.std.effect <- function(data, indices) {
  d <- data[indices,]
  0.642*smd(
    Mean.1 = mean(d$Achieve[d$Mex == 0], tr = 0.2),
    Mean.2 = mean(d$Achieve[d$Mex == 1], tr = 0.2),
    s.1 = sqrt(winvar(d$Achieve[d$Mex == 0], tr = 0.2)),
    s.2 = sqrt(winvar(d$Achieve[d$Mex == 1], tr = 0.2)),
    n.1 = length(d$Achieve[d$Mex == 0]),
    n.2 = length(d$Achieve[d$Mex == 1])
    )
  }

## Nonparametric bootstrap of robust standardized effect
> robust.boot <- boot(data = latino, statistic =
    robust.std.effect, R = 4999, strata = latino$Mex)

## Density plot of the bootstrap distribution
> plot(density(robust.boot$t))

## Mean of the bootstrapped effect size
> mean(robust.boot$t)
[1] 0.378868

## Standard error for the bootstrapped effect size
> sd(robust.boot$t)
[1] 0.1824919

## Obtaining the bootstrap limits
> boot.ci(boot.out = robust.boot, type = "bca")
BOOTSTRAP CONFIDENCE INTERVAL CALCULATIONS
Based on 4999 bootstrap replicates

CALL :
boot.ci(boot.out = std.boot, type = "bca")

Intervals :
Level        BCa
95%    ( 0.0089,  0.7230 )
Calculations and Intervals on Original Scale
```

Figure 9.6 shows a plot of the kernel density for the 4999 bootstrapped measures of robust standardized effect. This plot suggests that the bootstrap distribution is approximately normally distributed about the mean of 0.38. The SE of 0.18 suggests that there is a fair amount of variation in the bootstrap distribution.

9.7 SUMMARIZING THE RESULTS

Both point estimates and interval estimates of effect size should be reported in any write-up that includes results of hypothesis tests. It is up to the researcher whether she reports the standardized or unstandardized estimates. The write up for the analysis of the Latino data might include the following.

> **Sample Write-Up**
>
> A robust standardized effect size measure due to Algina, Keselman, and Penfield (2005) was computed yielding the value of 0.38. A nonparametric bootstrap bias-corrected-and accelerated 95% percentile interval using 4999 replicates was also computed for this effect yielding the range of $[0.01, 0.72]$. This range of values indicates likely candidates for the population standardized mean difference in educational achievement between Mexican and non-Mexican immigrants.

9.8 EXTENSION: BOOTSTRAPPING THE CONFIDENCE ENVELOPE FOR A Q-Q PLOT

A quantile-quantile (Q-Q) plot is often used as a graphical assessment of how far empirical data deviate from a particular probability distribution. This graph plots the empirical quantiles of the sample data versus the theoretical quantiles one would expect from a normal distribution. Often a reference line is added to the Q-Q plot to show what would be expected if the empirical and theoretical quantiles were equivalent. Command Snippet 9.9 shows the syntax to create a Q-Q plot of the standardized Mexican achievement scores using the qqnorm() function. A line is also added using the abline() function. The arguments a= and b= indicate the y-intercept and slope of the line to be plotted, respectively. Using a y-intercept of 0 and a slope of 1 will produce the reference line $y=x$.

Figure 9.7 shows the Q-Q plot of the standardized Mexican achievement scores. The observed standardized scores show deviation from the reference line indicating that these scores are not perfectly consistent with a normal distribution. Remember sample data are *never* normally distributed; the best one can hope for is consistency with such a distribution. Figure 9.8 shows the density plot of the empirical standardized Mexican achievement scores overlaid on the normal distribution. This plot and the Q-Q plot both show the same deviation between the empirical and theoretical distributions.

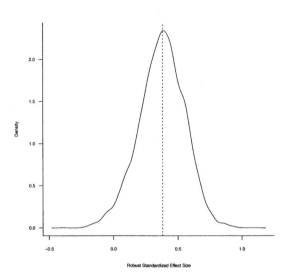

Figure 9.6: Density plot of the 4999 bootstrapped robust standardized effect sizes. The resampling was carried out under the alternative model, thus centering the distribution around the observed robust standardized effect of 0.38 (vertical dashed line).

Command Snippet 9.9: Syntax to create a Q-Q plot of the standardized Mexican achievement scores.

```
## Create a vector of standardized Mexican achievement scores
mex.z <- latino$z.achieve[latino$Mex == 1]

## Create the Q-Q plot
> qqnorm(mex.z, ylab = "Empirical Quantiles", xlab = "Normal
    Quantiles")

## Add the reference line
> abline(a = 0, b = 1)
```

9.9   CONFIDENCE ENVELOPES

Often researchers use graphical assessments to make decisions about whether the empirical data were drawn from a population that follows some particular probability distribution, such as the normal distribution. This is very useful when evaluating whether parametric assumptions of a model are plausible. In making this assessment, some amount of deviation is expected between the empirical data and theoretical models, since deviation occurs simply because of sampling variation. The key question is whether the deviation is within the expected variation due to sampling.

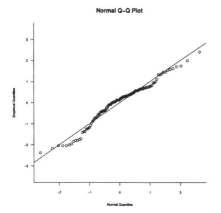

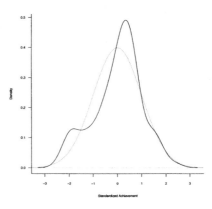

Figure 9.7: Q-Q plot of the standardized Mexican achievement scores. The line references where the points would lie if the sample of scores were normally distributed.

Figure 9.8: Density plot of the empirical standardized Mexican achievement scores. The normal probability model is shown with the dashed line.

A common approach to visualizing the expected variation from the reference line is to superimpose a *confidence envelope* on the Q-Q plot. A confidence envelope is essentially an interval estimate for a line or curve. It is called an envelope, since it graphically shows the degree of sampling error by placing an "envelope" around the reference line. To create this envelope, a *parametric bootstrap* is used. In the parametric bootstrap, replicate data sets are drawn from a theoretical distribution, such as the normal distribution, rather than from the observed data. Consider the process for using the parametric bootstrap to plot a confidence envelope for the Q-Q plot of the standardized Mexican achievement scores.

- Resample, with replacement, 116 observations from a standard normal distribution.

- Plot these observations versus the quantiles from the theoretical distribution.

- Repeat this process many times.

Since the replicate data sets are drawn from a known distribution—the standard normal—the Q-Q plot of the bootstrapped observations will deviate from the reference line only because of sampling error. By continually superimposing each of the bootstrapped Q-Q plots, one can visually see the expected variability from the reference line, which will aid assessment of the initial normality assumption.

Recall from Chapter 7 that the boot() function implements a parametric bootstrap using the argument sim="parametric" and ran.gen=, which takes a user-written function that describes how the bootstrapped observations are being generated. Command Snippet 9.10 shows a function used to sample observations from the standard normal distribution. The argument n=length(data) will compute the length of the vector provided in the data= argument of the boot() function and then sample that number of observations from the standard normal distribution. In this example 116 observations (the sample size of the Mexican group) will be sampled.

Command Snippet 9.10: A function to sample observations from the standard normal distribution.

```
> mex.gen <- function(data, mle){
  rnorm(n = length(data), mean = 0, sd = 1)
  }
```

Command Snippet 9.11 shows the implementation of the parametric bootstrap. Each of the 4999 bootstrap replicates contained in mex.qqboot has 116 resampled scores drawn from the standard normal distribution. Furthermore, the bootstrapped scores in each replicate are sorted in order from smallest to largest using the sort() function.

Command Snippet 9.11: The use of the boot() function to perform a parametric bootstrap.

```
> mex.qqboot <- boot(data = mex.z, statistic = sort, R = 4999,
    sim = "parametric", ran.gen = mex.gen)
```

Figure 9.9 shows a plot of the bootstrapped scores drawn in the first replicate versus the theoretical quantiles based on the standard normal distribution. The distribution of the bootstrapped observations shows very good agreement with the theoretical distribution. This is expected since the bootstrapped observations were resampled from a standard normal distribution.

Note that in Figure 9.9, while there is generally good agreement, there is still deviation from the reference line. Since the bootstrapped observations were drawn from a known standard normal distribution, this deviation is due only to sampling variation. To visually depict the expected deviation from the reference line due to sampling, many of the bootstrapped replicates are plotted versus the theoretical quantiles. Figure 9.10 shows the results of plotting 25 of these replicates.

Figure 9.10 visually shows the variation from the reference line. The envelope() function in the **boot** package is a quicker way in which to determine the limits for the confidence envelope. This function takes as its only required argument the results from the parametric bootstrap. Command Snippet 9.12 shows the use of this function to obtain the limits on the confidence envelope. These results are assigned

to an object, mex.env, and the structure of the object is examined using the str()
function.

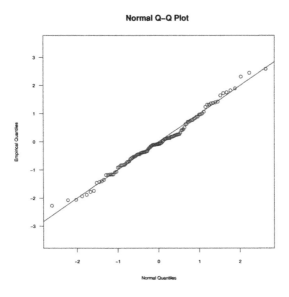

Figure 9.9: Q-Q plot of the bootstrapped scores drawn in the first replicate. The line
references where the points would lie if the sample of scores was consistent with a
normal distribution.

Command Snippet 9.12: The use of the envelope() function to obtain the limits
on the confidence envelope.

```
## This may produce an error message which can be ignored
> mex.env <- envelope(boot.out = mex.qqboot)

## Examine the structure of the bootstrap object
> str(mex.env)
List of 7
 $ point  : num [1:2, 1:116] -1.87 -3.56 -1.68 -2.86 -1.54 ...
 $ overall: num [1:2, 1:116] -1.58 -4.43 -1.42 -3.41 -1.32 ...
 $ k.pt   : num [1:2] 125 4875
 $ err.pt : num [1:2] 0.05 0.569
 $ k.ov   : num [1:2] 4 4996
 $ err.ov : num [1:2] 0.0016 0.0422
 $ err.nom: num [1:2] 0.05 0.05
```

The limits for the 95% simultaneous envelope are provided in the component
called overall in the bootstrap envelope object mex.env. This component is a
2×116 matrix that contains the upper limits of the envelope in the first row and

the lower limits of the envelope in the second row. In Command Snippet 9.13 the lines() function (see Chapter 4) is used to plot the limits of the confidence envelope.

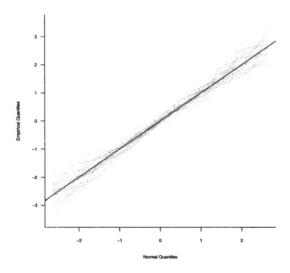

Figure 9.10: Q-Q plot of the bootstrapped standardized scores for $R = 25$ replicates of $n = 116$ resampled observations from the standard normal distribution. The bold line references where the points would lie if the sample of scores was consistent with a normal distribution.

The plot is shown in Figure 9.11. As seen in the figure, the points deviate from the reference line, although it is not more than one would expect because of sampling variation alone—most are inside of the confidence envelope. This is evidence that the distribution of the observed standardized achievement scores for Mexican immigrants could potentially have been drawn from a normally distributed population. Figure 9.12 shows the density plot for the standardized achievement scores along with a confidence envelope for the density plot. This was introduced in Section 3.4. These two plots both show the same pattern in the consistency with the normal model.

9.10   FURTHER READING

The bootstrap method for obtaining confidence intervals was introduced by Efron (1979). Efron (1981b) initally published on the methods for adjusting the confidence limits through studentizing and bias correction adjustment. Since that time many other methods for improving confidence limits using bootstrap methods have been proposed, including the approximate bootstrap confidence interval method (ABC; DiCicco & Efron, 1992) for dealing with multiparameter estimation; the use of calibration for improving coverage accuracy (Hall, 1986; Beran, 1987; Loh, 1987; Hall & Martin, 1988); the use of acceleration for bias correction (Efron, 1987; DiCicco, 1984); and the use of tilting to overcome difficulties associated with the BC_a estimation (Efron, 1981b).

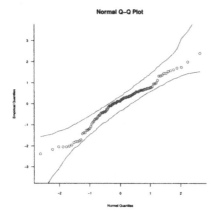
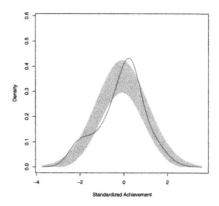

Figure 9.11: Q-Q plot of the studentized Mexican achievement scores along with a 95% confidence envelope. The confidence envelope was computed using $R = 4999$ replicates of $n = 116$ observations resampled using a parametric bootstrap from the $N(0, 1)$ distribution.

Figure 9.12: Density plot of the empirical standardized Mexican achievement scores. The confidence envelope based on the normal probability model is also added to the plot.

Command Snippet 9.13: R syntax to plot the Q-Q plot of the Mexican achievement scores along with the 95% confidence envelope.

```
## Plot the empirical versus the theoretical quantiles
## for the sorted Mexican standardized achievement scores
## and assign them to my.qq
> my.qq <- qqnorm(y = sort(mex.z), ylim = c(-3.5, 3.5), ylab =
    "Empirical Quantiles", xlab = "Normal Quantiles")

## Add the reference line
> abline(a = 0, b = 1, col = "red")

## Add the envelope's upper limits
> lines(x = my.qq$x, y = mex.env$overall[1, ])

## Add the envelope's lower limits
> lines(x = my.qq$x, y = mex.env$overall[2, ])
```

Several studies have been undertaken to empirically compare these methods (Hall, 1988; Lee & Young, 1995; DiCicco & Efron, 1996). The last of these—a survey of the bootstrap methodology at the time—was followed by four articles in which authors commented on the bootstrap methodology. Carpenter and Bithell (2000) wrote an especially readable treatise on many of the bootstrap interval methods. Finally, it should be pointed out that Rubin (1981) proposed a methodology for bootstrapping that is Bayesian in nature.

PROBLEMS

9.1   Institutional data from 30 Minnesota colleges and universities were obtained from 2006 and can be found in the *MNColleges.csv* data set. Among the variables collected were expenditures per student, Expend (in U.S. dollars), and whether the college or university is a private or public institution, Sector. Of interest is whether expenditures differ by sector.

a) Conduct a nonparametric bootstrap test of the mean difference.

b) Compute the bootstrap interval using the bias-corrected-and-accelerated (BC_a) method.

c) Compute the studentized bootstrap interval.

d) Compare and contrast the two intervals.

e) Compute a point estimate of the standardized effect size for the mean difference in expenditures by sector. Should a robust effect size be alternatively used? Explain.

f) Based on your answer to the previous question, provide an interval estimate for the true population standardized effect.

g) Write up the results from all analyses, in no more than three paragraphs, as if you were writing a manuscript for publication in a journal in your substantive area.

CHAPTER 10

DEPENDENT SAMPLES

Statistical independence is far too often assumed casually, without serious concern for how common is dependence and how difficult it can be to achieve independence (or related structures).

—W. Kruskal (1988)

In Chapter 7, the method of the bootstrap was introduced to examine group differences when the assumption of independence has been met. However, in many cases, this assumption is not met. Sometimes the study design employed by the researcher leads to cases where the two samples are dependent. For example, research study designs that use matching, as well as within-subjects research designs (e.g., pretest–posttest designs) both induce dependence among the observations. In such cases, the analytic methods for inference introduced thus far would be inappropriate. In this chapter methods are presented that can be used to analyze group differences when the two samples are not independent.

Comparing Groups: Randomization and Bootstrap Methods Using R **207**
First Edition. By Andrew S. Zieffler, Jeffrey R. Harring, & Jeffrey D. Long
Copyright © 2011 John Wiley & Sons, Inc.

10.1   MATCHING: REDUCING THE LIKELIHOOD OF NONEQUIVALENT GROUPS

Recall from Chapter 6, that the specification of a comparison group is an essential step in the design of good educational research. Recall further, that the best comparison group is one that is "composed of people who are similar to the people in the treatment group in all ways except that they did not receive the treatment" (Light et al., 1990, p. 106). The best way to achieve this is through random assignment of the study participants to treatment and control groups. In Chapter 6, random assigment of study participants to conditions (or conditions to participants) was described as a method that ensures that the treatment group and the comparison group are equivalent, *on average*, on all attributes, characteristics, and variables other than the treatment.

While this is a true description of what should happen with random assignment, what happens in practice may not meet this description. Many researchers find that even with random assignment, their two groups differ on average—even reliably—on certain characteristics or measures. This may not be a failure of the random assignment to produce groups that are, on average, equivalent. Rather, these differences may be "due to chance" and are to be expected, especially when the sample size is not large. Across several studies it is expected that the differences would balance out between the groups. This is the true description of equivalent groups "on average."

One method of ensuring that gross imbalances among the groups does not occur is to *match* samples using variables that relate to the outcome under study, or *covariates*. When matching, researchers group study participants with similar scores on a covariate *prior to the random assignment*. Every pair of matched participants—which are referred to as a block—are then split after matching, and one is randomly assigned to the treatment group and the other to the control group.[1] The idea of matching is that the researcher can create control and treatment groups that are more alike—at least on the covariate(s) used in the matching—thereby reducing the probability that imbalance will occur.

10.2   MATHEMATICS ACHIEVEMENT STUDY DESIGN

Consider the following study in which the researchers are interested in whether a particular curricular approach is effective for increasing the mathematics scores for at-risk students. The researchers have 40 at-risk students who have volunteered to take part in a summer program during which the curricular approach to be evaluated will be compared to the older approach. The researchers intend to randomly assign these 40 students to each of the two curricula. Because of the small sample size, however, the researchers are a little nervous that simple random assignment to the control and treatment curricula may result in imbalance between the two groups.

---

[1]Some authors and textbooks distinguish between blocking—using groups with similar scores—and matching—using groups with the exact same scores. In this monograph the terms are used interchangeably. It is also possible to have more than two groups by creating blocks using more participants.

They have decided to match students based on their PSAT mathematics score—a variable they believe should be related to their outcome variable.

The data set *PSAT.csv* contains data on the 40 at-risk students' PSAT scores along with their participant ID number. Command Snippet 10.1 shows the syntax to read in the *PSAT.csv* data and examine the data frame object.

Command Snippet 10.1: Syntax to read in the *PSAT.csv* data and the use of the order() function.

```
## Read in the data
> psat <- read.table(file = "/Documents/Data/PSAT.csv", sep =
    ",", header = TRUE, row.names = "ID")

## Examine the data frame object
## Output is suppressed
> head(psat)
> tail(psat)
> str(psat)
> summary(psat)
```

After reading this data into R, the sort() function can be used to order the participants' PSAT scores. This function takes an argument x= to provide a vector on which the sort() function will order the scores. By default, the vector is sorted in ascending order. The optional argument decreasing=TRUE can be provided to sort in descending order. Command Snippet 10.2 shows the syntax to arrange the participants' PSAT scores in ascending order.

Command Snippet 10.2: Syntax to use of the sort() function to arrange the participants' PSAT scores in ascending order.

```
#### Read in the data
> sort(math$PSAT)
 [1] 270 275 335 360 400 405 415 415 450 460 475 475 480
[14] 480 485 485 490 490 495 495 500 510 530 535 540 540
[27] 545 545 550 550 575 585 630 635 640 645 680 700 780
[40] 795
```

Since there are two treatments in the study, blocks can be formed by pairing the two study participants with the highest PSAT scores, the next two highest PSAT scores, etc. The random assignment is then carried out within each block. Thus for the first block, either the participant with a PSAT score of 270 (participant 19) or the participant with a PSAT score of 275 (participant 8) is randomly assigned to treatment and the other to control. Likewise, one of the participants from the second block (the participant with a PSAT score of 335 and the participant with a PSAT score of 360) is randomly assigned to treatment and the other to control. The process is continued in this manner until all of the study participants have been assigned.

The data set *OrderedPSAT.csv* contains the PSAT scores for the 40 participants. There is also a variable indicating the condition that the participant was assigned to. Command Snippet 10.3 shows the syntax to read in the *OrderedPSAT.csv* data and examine the data frame object.

Command Snippet 10.3: Code to read in the *OrderedPSAT.csv* data and examine the data frame object.

```
## Read in the OrderedPSAT.csv data
> psat2 <- read.table(file = "/Documents/Data/OrderedPSAT.csv",
    sep = ",", header = TRUE)

## Make sure the data read in correctly
## Output is suppressed
> head(psat2)
> tail(psat2)
> str(psat2)
> summary(psat2)
```

To determine whether the matching achieved the desired goal of balancing the two groups—at least on the covariate of PSAT score—the distribution of the covariate for each group is examined. Command Snippet 10.4 shows the syntax used to examine the PSAT scores for the two treatments. Figure 10.1 shows the density plots of the PSAT scores for both the control and treatment conditions. These plots suggest that the two groups are very similar in terms of their PSAT scores. Numerical summaries of these distributions also suggest that the PSAT scores for the control group ($M = 516, \text{SD} = 116$) and the treatment group ($M = 517, \text{SD} = 118$) are similar.

Command Snippet 10.4: Syntax to examine the distribution of PSAT scores conditional on treatment condition.

```
## Plot the PSAT scores for each condition
> plot(density(psat2$PSAT[psat2$Condition == "Control"]), lty =
    "dashed")
> lines(density(psat2$PSAT[psat2$Condition == "Treatment"]),
    lty = "solid")

## Mean of PSAT scores
>   tapply(X = psat2$PSAT, INDEX = psat2$Condition, FUN = mean)
  Control Treatment
   516.75    515.50

## Standard deviations of the PSAT scores
>   tapply(X = psat2$PSAT, INDEX = psat2$Condition, FUN = sd)
  Control Treatment
  116.4189  117.7967
```

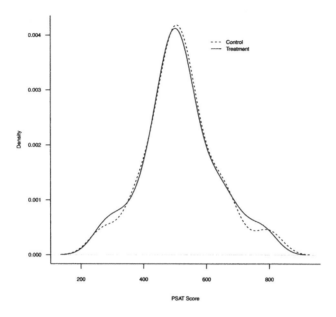

Figure 10.1: Distribution of PSAT scores conditioned on treatment.

These analyses suggest that the matching was effective in balancing the two groups—on PSAT scores—prior to receiving the curricula.

10.2.1 Exploratory Analysis

After being exposed to either the control curriculum or the treatment curriculum, all students were given a mathematics achievement test. The data set *BlockedPSAT.csv* contains the PSAT scores for the 40 participants, the condition each participant was assigned to, and the participants' mathematics achievement scores. Command Snippet 10.5 shows the syntax used to read in these data, examine the data frame object, and carry out the exploratory analysis on the mathematics achievement scores.

The density plots (see Figure 10.2) and side-by-side box-and-whiskers plots (see Figure 10.3) suggest that the distribution of students' mathematics achievement scores for both curricula are negatively skewed. They also suggest that there is comparable variation in these achievement scores between the two curricula. Finally, there is a difference in location for these two distributions, with students who experienced the treatment curriculum showing both a higher median and mean mathematics achievement score than students who experienced the control curriculum.

While the sample distributions suggest a difference in the average mathematics achievement scores, again the broader question is asked of whether or not this

Command Snippet 10.5: Syntax to examine the distribution of mathematics achievement scores conditional on treatment condition.

```
## Read in the BlockedPSAT.csv data
> math <- read.table(file = "/Documents/Data/BlockedPSAT.csv",
    sep = ",", header = TRUE)

## Make sure the data read in correctly
## Output is suppressed
> head(math)
> tail(math)
> str(math)
> summary(math)

## Density plots
> plot(density(math$Achievement[math$Condition == "Control"],
    bw = 4), lty = "dashed")
> lines(density(math$Achievement[math$Condition ==
    "Treatment"], bw = 4), lty = "solid" )

## Side-by-side box-and-whiskers plots
> boxplot(math$Achievement[math$Condition == "Control"],
    math$Achievement[math$Condition == "Treatment"], names =
    c("Control", "Treatment"))

## Conditional means
> tapply(X = math$Achievement, INDEX = math$Condition, FUN =
    mean)
  Control  Treatment
   26.55      29.35

## Conditional standard deviations
> tapply(X = math$Achievement, INDEX = math$Condition, FUN =
    sd)
  Control  Treatment
 5.529585  6.983251
```

observed sample mean difference of 2.8 achievement points is just an artifact of the random process used to assign students to treatments, or if it is a "real" effect reflecting population group differences. To examine this, the null hypothesis,

H_0 : The new curricular approach is no more effective than the older approach,

is tested.

In order to test this hypothesis, a randomization test is carried out. This is a test of the expected variation in the mean difference under the assumption of no effect. However, because these data come from study participants who have been matched prior to the employment of random assignment, the assumption of exchangeability required for the randomization method is no longer appropriate.

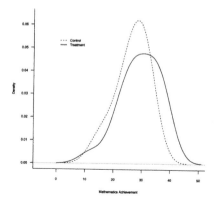

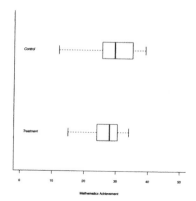

Figure 10.2: Density plots for the distribution of mathematics achievement scores conditioned on treatment.

Figure 10.3: Side-by-side box-and-whiskers plots for the distribution of mathematics achievement scores conditioned on treatment.

These data are arranged based on the matching and assignment that was carried out on the PSAT scores. Note that in this data set each row, or case, is a block rather than an individual student. This is a common format for organizing matched data so that the matched participants are linked via the row.

10.3 RANDOMIZATION/PERMUTATION TEST FOR DEPENDENT SAMPLES

When researchers have matched study participants and then randomly assigned them to treatment and control within those blocks, an appropriate analysis to determine whether there are statistically reliable differences between the groups is the randomization test. Recall that in order to obtain the reference distribution in which to evaluate the observed mean difference, the null hypothesis is assumed to be true, and the data are randomly permuted in a manner that is consistent with the initial design assumed to give rise to the observed data.

For independent samples, all the observations were put into one "supergroup" as the null hypothesis of no difference was assumed to be true. Then the observations were randomly permuted to treatment and control groups consistent with the original design. For a design that uses matching, the random assignment takes place only within each block. Thus the supergroup is not all observations, but rather a block. This is consistent with the null hypothesis, which suggests that there is no difference in groups *after matching on PSAT scores*.

Because the null hypothesis says that the two populations are identical only after matching on PSAT scores, the way in which the data is permuted needs to be modified for this design. If the assumption of the null hypothesis of no difference is true, the treatment and control labels would have no meaning only after matching on PSAT scores. Thus, permuting the data under this hypothesis randomly assigns scores to "treatment" and "control" only after the matching has taken place. Thus, the data are randomly permuted with the caveat that observations of a particular block are assigned to different groups.

After forming the treatment and control scores by randomly permuting the observations in each block, one could calculate the mean difference in mathematics achievement for the two permuted groups. If the sample size is small enough, all possible mean differences could easily be produced by just listing out all of the possibilities. The observed mean difference can then be evaluated in that exact distribution. Instead, a Monte Carlo method is used to simulate the exact distribution and compute a simulated p-value.

10.3.1 Reshaping the Data

The `math` data frame is arranged so that each row contains the data for a single participant in the study. This representation, referred to as the *long format*, is the arrangement that has been used thus far in the monograph. Matched data are often presented in the *wide format*. The wide format typically arranges the data so that each row corresponds to a block, rather than a participant. Tables 10.1 and 10.2 show the data for the participants in the first two blocks in both the long and wide formats, respectively.

Table 10.1: Data for Participants in First Two Blocks Presented in Long Format

Block	PSAT	Condition	Achievement
1	270	Control	15
1	275	Treatment	27
2	335	Control	15
2	360	Treatment	12

The `reshape()` function is used to rearrange the data between the two formats. In this function, the `data=` argument is used to specify the data frame that will be rearranged. The `direction=` argument takes a character string of either `"wide"` or `"long"`, depending on the format the data should take. The `idvar=` argument indicates the variable(s) in the long formatted data that identify multiple records from the same block, group, or individual. The `timevar=` argument specifies the variable in the long formatted data that differentiates multiple records from the same block, group, or individual. Lastly, the argument `v.names=` indicates which of the variables in the long formatted data are to be made into multiple variables when the data is rearranged into the wide format.

Table 10.2: Data for Participants in First Two Blocks Presented in Wide Format

	Control		Treatment	
Block	**PSAT**	**Achievement**	**PSAT**	**Achievement**
1	270	15	275	27
2	335	15	360	12

For example, to rearrange long formatted data from Table 10.1 to the wide format in Table 10.2 the argument direction="wide" is provided in the reshape() function. The Block variable identifies the records that are in the same block, so the argument idvar="Block" is used. The argument timevar="Condition" is used since the Condition variable differentiates the two conditions for each block. Finally, v.names = c("PSAT", "Math") is provided to create multiple variables—one for control and one for treatment—out of both the PSAT and Achievement variables. Command Snippet 10.6 shows the complete syntax to reshape the math data frame.

Command Snippet 10.6: Syntax to reshape the math achievement data.

```
## Reshape the math achievement data to the wide format
> math.wide <- reshape(data = math, direction = "wide", idvar =
    "Block", timevar = "Condition", v.names = c("PSAT",
    "Achievement"))

## Examine the first part of the data frame object
> head(math.wide)
  Block PSAT.Control Achievement.Control
1     1          270                  15
3     2          360                  15
5     3          400                  24
7     4          415                  24
9     5          460                  26
11    6          475                  19
   PSAT.Treatment Achievement.Treatment
1             275                     27
3             335                     12
5             405                     22
7             415                     27
9             450                     20
11            475                     27
```

Before carrying out the randomization, the data frame is simplified by only considering the data in the Achievement.Control and Achievement.Treatment columns. These columns are stored in a new data frame object called math2. Command Snippet 10.7 shows the syntax to create this new data frame object.

Command Snippet 10.7: Syntax to create a new data frame object containing only the math achievement scores for the control and treatment groups.

```
> math2 <- math.wide[ ,c(3, 5)]
```

10.3.2   Randomization Test Using the Reshaped Data

In terms of the data frame, the problem becomes one of randomly permuting the observations within each row, as they are the blocks. To permute the data the sample() function is used. Recall from Chapter 6 that the apply() function is used to apply a particular function to the rows or columns of a matrix or data frame. Since the observations in a block need to be randomly assigned to groups, MARGIN=1 is used. The argument FUN=sample applies the sample() function to each row. Command Snippet 10.8 shows the syntax for sampling within rows of the math data frame.

Command Snippet 10.8: Syntax to sample within each row of the math data frame.

```
> apply(X = math2, MARGIN = 1, FUN = sample)
      1   3   5   7   9  11  13  15  17  19  21  23  25  27  29  31
[1,] 15  12  24  27  20  27  25  28  32  27  34  32  31  31  38  34
[2,] 27  15  22  24  26  19  23  22  27  24  25  30  38  33  29  32
     33  35  37  39
[1,] 32  39  29  36
[2,] 34  29  36  30
```

The result from using the apply() function is a matrix with 2 rows and 20 columns. The first row is arbitrarily designated the control group and the second row, the treatment group. To find the mean for each row, the apply() function is again used—this time changing the FUN= argument to apply the mean() function. Command Snippet 10.9 shows the syntax to perform the permutation *and* compute the mean for both groups' permuted observations.

Command Snippet 10.9: Syntax to sample within each row of the math data frame and then compute the mean for each row in the resulting matrix.

```
> apply(X = apply(X = math2, MARGIN = 1, FUN = sample), MARGIN
    = 1, FUN = mean)
[1] 28.65 27.25
```

To compute the difference between these two means, the diff() function can be used. Command Snippet 10.10 shows the syntax to perform the permutation, compute the mean for both groups' permuted observations, and compute the difference between those means.

Command Snippet 10.10: Syntax to sample within each row of the math data frame, compute the mean for each row in the resulting matrix, and compute the mean difference.

```
> diff(apply(X = apply(X = math2, MARGIN = 1, FUN = sample),
    MARGIN = 1, FUN = mean))
[1] -1.4
```

The `replicate()` function is now used to carry out many permutations of the data, each time computing the difference between the permuted groups' means. Command Snippet 10.11 shows the syntax for carrying out this computation. The randomization distribution is also plotted and examined. Finally, the observed mean difference of 2 is evaluated in the randomization distribution.

Command Snippet 10.11: Syntax to carry out a Monte Carlo simulation of the randomization test for the matched achievement data.

```
## Carry out 4999 permutations of the data
> permuted <- replicate(n = 4999, expr = diff(apply(X = apply(X
    = math2, MARGIN = 1, FUN = sample), MARGIN = 1, FUN =
    mean)))

## Plot the randomization distribution
> plot(density(permuted), xlab = "Permuted Mean Difference",
    main = " ")

## Compute the mean of the randomization distribution
> mean(permuted)
[1] -0.001380276

## Compute the standard error of the randomization distribution
> sd(permuted)
[1] 1.355299

## Count the number of permuted mean differences
## as extreme or more extreme than 2.8
> length(permuted[abs(permuted) >= 2.8])
[1] 211

## Compute p-value
> (211 + 1) / (4999 + 1)
[1] 0.0424
```

The Monte Carlo randomization distribution (shown in Figure 10.4 is approximately normal with mean of 0 and an estimated standard error of 1.4. This indicates that under the assumption of no difference of achievement scores between the two treatments, that the variation expected in the mean difference simply because of random assignment is quite small. It is expected that most of the mean differences to be within ± 2.8 achievement points from zero based on these 4999 permutations of the data. Of the 4999 permuted mean differences, 211 were as extreme or more extreme than the observed sample difference of 2.8. Using the correction for the Monte Carlo simulation, the Monte Carlo p-value for the example would be

$$p = \frac{211 + 1}{4999 + 1} = 0.042. \qquad (10.1)$$

This p-value shows moderate evidence against the null hypothesis of no effect of the new curricular approach on mathematics achievement.

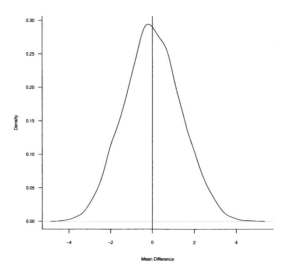

Figure 10.4: Randomization distribution of the mean difference under the null hypothesis of no difference. A vertical line (solid) is drawn at 0, the hypothesized mean difference.

10.4 EFFECT SIZE

The effect size for two dependent samples is computed quite similarly to that for independent samples. Of interest is finding out how different the mean of the treatment group is from the mean of the control group in the population. This can be estimated from the sample means. Since the raw scores are difficult to interpret, the metric is standardized by dividing this difference by the standard deviation of the control groups' achievement scores. This standardized measure of effect is called Glass's delta. Glass (1977) argued that when a study has employed an actual control group, by using the standard deviation of the control group, rather than the pooled estimate introduced in Chapter 9 in the denominator, the standardized effect would not change due to differences in variation when several treatment options are being compared. In general this Glass's delta is computed as

$$\Delta = \frac{|\mu_{\text{Treatment}} - \mu_{\text{Control}}|}{\sigma_{\text{Control}}}, \tag{10.2}$$

which can be estimated using the sample data in this example as

$$\hat{\Delta} = \frac{|29.35 - 26.55|}{5.53} = 0.51. \tag{10.3}$$

This indicates that on average, the scores in the experimental and control conditions differ by 0.51 standard deviations. It is also possible to compute an interval estimate for the standardized mean difference using block bootstrapping (see Section 10.7).

10.5 SUMMARIZING THE RESULTS OF A DEPENDENT SAMPLES TEST FOR PUBLICATION

An example write-up for the randomization test results is given below. As with any simulation method, it is important to report the method used to analyze the data along with the p-value. It is also necessary to report whether the p-value is exact or approximated via Monte Carlo methods. If the latter is the case, you should indicate the number of data permutations that were carried out and whether the correction for the p-value was used.

> **Sample Write-Up**
>
> Forty at-risk high school students were matched on their PSAT scores and then randomly assigned to an intensive summer training in mathematics ($n = 20$) or were given "treatment as usual" ($n = 20$). Treatment as usual consisted of the typical summer curriculum the students previously experienced. Sample differences in the two groups suggested that students who received the intensive summer curriculum ($M = 28, \text{SD} = 6$) scored, on average, 2 achievement points higher than students who received the typical summer curriculum ($M = 26, \text{SD} = 6$). A randomization test was used to determine whether this difference was statistically reliable. A Monte Carlo p-value was computed by permuting the data 4999 times. Using a correction suggested by Davison and Hinkley (1997), a p-value of 0.042 was computed. This is moderate evidence (see Efron & Gous, 2001) against the null hypothesis of no difference. The standardized mean difference as measured by an estimate of Glass's delta was 0.51. A nonparametric block bootstrap 95% percentile interval using 4999 replications was also computed for this effect as $[0.13, 1.12]$.

10.6 TO MATCH OR NOT TO MATCH ... THAT IS THE QUESTION

Consider what would happen had the assumption of independence incorrectly been made. Then the permutations of the data would be carried out under random assignment of all the data rather than within blocks. Command Snippet 10.12 shows the syntax for carrying out the Monte Carlo randomization under this incorrect assumption.

Command Snippet 10.12: Using the math data to carry out a Monte Carlo randomization test under the incorrect assumption of independence.

```
## Put all scores in one common vector
> all.math <- c(math2$Achievement.Control,
    math2$Achievement.Treatment)

## Obtain 4999 permutations of the data
> permuted.ind <- replicate (n = 4999, expr = sample(all.math))

## Write a function to compute the mean difference
> mean.diff <- function (data) {
  mean(data[1:20]) - mean(data[21:40])
  }

## Compute the mean difference for each vector of permuted data
> diffs <- apply(X = permuted.ind , MARGIN = 2, FUN =
    mean.diff)

## Count extreme values
> length(diffs[abs(diffs) >= 2.8])
[1] 928

##Compute the p-value
> (928 + 1) / (4999 + 1)
[1] 0.1858
```

The p-value based on the method carried out under the assumption of independence is 0.186. This would suggest that the data show very weak evidence against the null hypothesis of no effect of the new curriculum. This inference is very different than the prior inference from the randomization method ($p = 0.042$). An important thing to remember is that *the validity of the inferences you draw are directly reliant on whether or not the observed data meets the assumptions of the method you are using.*

Why does the p-value change this dramatically? The size of the p-value is directly related to the variation in the distribution of the test statistic. In this case it is the distribution of the mean difference. Figure 10.5 shows the randomization distribution of the mean difference for both the assumptions of independence and dependence. Both of these randomization distributions are symmetric and centered at zero. The variation for the randomization distribution assuming independence is SE = 2.05. This is much higher than that for the distribution not assuming independence, SE = 1.36.

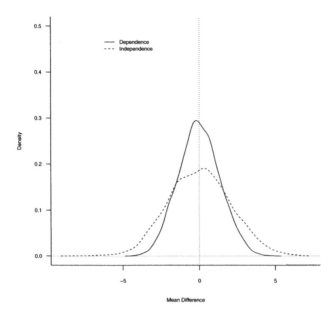

Figure 10.5: Randomization distribution of the mean difference for both the assumptions of independence and dependence. A vertical line (dotted) is drawn at the hypothesized mean difference of 0.

This implies that the mean difference varies much more assuming independence just because of random assignment. Because of this, the observed difference of 2.8, or a difference more extreme, is much less likely when the data are dependent than when they are independent.

The variation in the randomization distribution is affected by (1) the variation in the observed data, (2) the sample size, and (3) the number of permutations of the data carried out. The effect of the number of permutations on the variation was addressed in Chapter 6. Chapter 9 addressed how sample size affects this variation. The assumption of independence or dependence affects the first of these, namely the variation in the observed data.

When a researcher uses blocking, the blocks should be more homogeneous (less variable) in terms of the blocking variable than the whole set of participants. Since the blocking variable is chosen because of its relation to the outcome variable, this implies that there will also be less variation on the outcome variable within blocks. This reduction in variation directly impacts the variation in the randomization distribution.

If a design with matching is employed, it is essential that the researcher matches using covariate(s) that relate to the outcome being measured (see Shadish et al., 2002).

Because of this, a pretest or baseline measure of the outcome variable is often the best covariate on which to match study participants. If a pretest is not possible to obtain, the covariate should be as closely related to the outcome as possible. Matching on covariates unrelated to the outcome variable provide little benefit, as the variation on the outcome will not be reduced within blocks. This can actually lead to a reduction in the likelihood of finding evidence of an effect.

10.7   EXTENSION: BLOCK BOOTSTRAP

The block bootstrap is a nonparametric bootstrap method in which observations are drawn from the sample without replacement. Rather than resampling individual observations, however, this method resamples entire blocks of observations. By resampling blocks of observations, the dependent nature of the data is kept intact. Figure 10.6 shows the steps used to carry out the block bootstrap to obtain the distribution for a particular test statistic, $\hat{V}$, under the alternative model.

Nonparametric Block Bootstrapping

- Randomly resample n blocks from the observed sample data *with replacement* $\{b_1^*, b_1^*, b_2^*, b_3^*, \ldots, b_{n_1}^*\}$.

- Compute $\hat{V}$—the test statistic of interest—using the observations obtained from the bootstrapping.

- Repeat the first two steps many times, say R times, each time recording the statistic $\hat{V}$.

- The distribution of $\hat{V}_1, \hat{V}_2, \hat{V}_3, \ldots, \hat{V}_R$ can be used as an estimate of the sampling distribution of V under the alternative model.

Figure 10.6: The steps used to block bootstrap the distribution for a particular test statistic, $\hat{V}$, under the alternative model.

Consider how the bootstrap distribution would be created under the alternative model. The purpose is to estimate the standardized difference of the effect in mathematics achievement for the mathematics achievement data.

- First, resample 20 blocks of paired observations with replacement from the sample of mathematics achievement data. This is the bootstrap replicate data set.

- Calculate Glass's standardized effect for the bootstrap replicate data set.

- Repeat the process many times to obtain the bootstrap distribution of Glass's standardized effect.

To bootstrap under this alternative model in R, the **boot** package is used. The **boot** package implements block bootstrapping using the `tsboot()` function. This function requires the initial data to be a time-series object, which will suffice for any type of dependence among pairs of observations, as in the matching example. The `as.ts()` function is used in Command Snippet 10.13 to coerce the `math2` data frame into a time-series object.

Command Snippet 10.13: Coercing the `math2` data frame into a time-series object.

```
> ts.math <- as.ts(math2)
```

It is common to treat matched dependent data and time-series data in a similar manner. Since the study participants were matched prior to being randomly assigned to treatments, the two participants in the same block can be treated as the "same" participant who was measured pretreatment and then again posttreatment—as in a time series. A function is then written that will compute Glass's standardized effect. Command Snippet 10.14 includes a function called `glass.delta()` that when applied to the `ts.math` object will compute the Glass's delta between the mathematics achievement scores for the resampled treatment and control participants. Two things to note about this function are that the argument `indices` is not provided, and the variables are accessed via the column numbers rather than their names.

Command Snippet 10.14: A function to compute the standardized effect between the mathematics achievement scores for the resampled treatment and control participants. The columns in a time-series object need to be called through indexing. The function is also tested on the time-series data.

```
## Function to compute Glass's delta
> glass.delta <- function(data){
  numerator <- abs(mean(data[ ,2]) - mean(data[ ,1]))
  denominator <- sd(data[ ,1])
  numerator / denominator
  }

## Test the glass.delta() function
> glass.delta(ts.math)
[1] 0.5063671
```

After executing this function in R, it is tested by using it with the time-series data in `ts.math`. This result is within rounding of the value for Glass's delta obtained in Command Snippet 10.3. The `tsboot()` function is used to perform the actual bootstrapping. This function takes the arguments `tseries=` to input the name of the time-series object to be bootstrapped; `statistic=` to indicate the name of the function that determines what is to be computed with each replicate data set; R= to indicate how many replicate data sets should be resampled; sim="fixed" to indicate that block resampling is to be carried out; and l= to indicate the length of each block. Command Snippet 10.15 shows the syntax for using the tsboot() function to bootstrap the standardized difference in achievement scores for 4999 replicates. The results are assigned to an object called match.boot.

Command Snippet 10.15: Bootstrapping the mean difference in achievement scores for 4999 replicates. The bootstrapping is carried out within strata rather than by pooling as in the last chapter. These results are also plotted and displayed.

```
## Load the boot package
> library(boot)

## Carry out the block bootstrap
> match.boot <- tsboot(tseries = ts.math, statistic =
    glass.delta, R = 4999, sim = "fixed", l = 2)

## Plot the bootstrap distribution
> plot(density(match.boot$t))

## Mean of the bootstrap distribution
> mean(match.boot$t)
[1] 0.5548091

## Standard error of the bootstrap distribution
> sd(match.boot$t)
[1] 0.2510088

## Examine the bootstrap object
> match.boot

BLOCK BOOTSTRAP FOR TIME SERIES

Fixed Block Length of 2

Call:
tsboot(tseries = ts.math, statistic = glass.delta, R = 4999,
    l = 2, sim = "fixed")


Bootstrap Statistics :
     original    bias    std. error
t1* 0.5063671 0.048442   0.2510088
```

Figure 10.7 shows the plotted results of the bootstrapped Glass's delta under the alternative model. This distribution is centered at 0.55, which is higher than the observed effect of 0.51. This difference, which is quantified in the bias statistic, along with the asymmetry in the plot indicate that the estimated effect may be

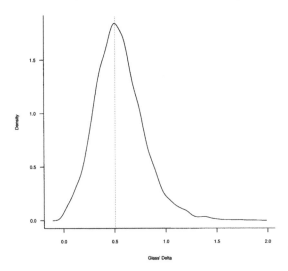

Figure 10.7: Bootstrap distribution of the standardized mean difference as measured by Glass's delta in mathematics achievement scores between treatment and control participants for 4999 replicates. The resampling was carried out using block bootstrapping under the alternative model. The vertical line (dotted) is drawn at the observed standardized mean difference of 0.51.

slightly positively biased (an overestimate of the population effect).[2] The `ci.boot()` function is used to obtain an interval estimate for this effect. Since bias-corrected-and-accelerated intervals cannot be computed for time-series data, the results based on the percentile interval are reported. The percentile interval limits can be obtained using the argument `type="perc"` as shown in Command Snippet 10.16. In this case, the interval of plausible standardized differences is $[0.13, 1.12]$.

10.8 FURTHER READING

The use of statistical matching in the design of research has a long history. Throughout this history, many different methods used to match study participants have been proposed and used including exact matching, caliper matching, index matching, cluster group matching, and optimal matching, to name a few. Matching on composites created from multiple variables can also be carried out. For further reading on these methods, see Cochran (1965), Cochran and Rubin (1973), Henry and MacMillan (1993), and Rosenbaum (1995).

---

[2] Using a more robust estimate of the standardized effect would result in a statistic that is less biased.

Command Snippet 10.16: Obtaining the limits for the nonparametric percentile bootstrap confidence interval for the standardized mean difference in mathematics achievement scores for 4999 replicates.

```
> boot.ci(match.boot, type = "perc")
BOOTSTRAP CONFIDENCE INTERVAL CALCULATIONS
Based on 4999 bootstrap replicates

CALL :
boot.ci(boot.out = match.boot, type = "perc")

Intervals :
Level       Percentile
95%    ( 0.1346,  1.1184 )
Calculations and Intervals on Original Scale
```

Current work using matching is being carried out in the development and application of propensity score methods (PSM). These methods use matching in an attempt to create equivalent groups in observational studies to strengthen the causal inferences that can be drawn. Rosenbaum and Rubin (1983) wrote the seminal paper on PSM. Some other good starting points to read more about PSM include Shadish et al. (2002), Oakes and Johnson (2006), and D'Agostino (1998).

PROBLEMS

The Brief Psychiatric Rating Scale (BPRS) assesses the level of symptom constructs (e.g., hostility, suspiciousness, hallucination, grandiosity, etc.) in patients who have moderate to severe psychoses. Each symptom is rated from 1 (not present) to 7 (extremely severe) and depending on the version between a total of 18 and 24 symptoms are scored. The data set *BPRS.csv* contains 2 measurements, a baseline and posttreatment measurement, based on the 18 symptoms scale for each of 40 men. The researchers who collected these data were interested in whether or not there is an effect of treatment on improving (lowering) the measured symptoms. Because it was felt that every participant would benefit from the treatment, a traditional control group could not be used. Rather, the researchers employed a pre–post design where each participant acts as her/his own control. In this sense, the measurements are considered matched—the pre–post measurements for each study participant can be considered a block. Use these data to answer the following problems.

10.1 Perform an exploratory data analysis of these data, using both numerical and graphical summaries, to determine whether or not there is an effect of treatment on improving (lowering) the measured symptoms. Write up the results of this analysis.

10.2 Perform an inferential analysis of this data. Use a randomization test to evaluate whether there is an effect of the treatment for the 40 male participants. In the analysis, compute a point estimate for the standardized effect. Use a nonparametric block bootstrap to compute the 95% percentile interval for this effect. Write up the results as if you were writing a manuscript for publication in a journal in your substantive area.

CHAPTER 11

PLANNED CONTRASTS

> *... contrast analysis permits us to ask focused questions of our data ... to ensure that the results are compared to the predictions we make based on theory, hypothesis, or hunch.*
>
> —R. Rosenthal & R. L. Rosnow (1985)

In Chapters 6–9 methods of inference for comparing two groups were presented. These methods can be considered as extensions of the exploratory analyses introduced in Chapter 4. In Chapter 5, the exploratory analysis focused on the comparison of more than two groups. In particular the research question of whether there were per capita expenditure differences between the seven regions of Vietnam was examined. In this chapter, methods of inference are presented in which the focus is the comparison of more than two groups.

When dealing with more than two groups, a useful distinction is between an *omnibus test* and *specific comparisons*. An omnibus test involves all the groups, and is used to evaluate the general null hypothesis that the group means are equal. A specific comparison typically involves a subset of the group—usually a pair—and evaluates the more specialized null hypothesis that the two given group means are equal. There is only one omnibus test, but usually many specific comparisons. The specific comparisons can be quite numerous if the number of groups is large (say,

Comparing Groups: Randomization and Bootstrap Methods Using R
First Edition. By Andrew S. Zieffler, Jeffrey R. Harring, & Jeffrey D. Long
Copyright © 2011 John Wiley & Sons, Inc.

10 or so). Because of the number of specific comparisons, some researchers and statisticians have been concerned with possible complications of conducting many tests. This has led to the development of the *adjustment procedures* discussed in Chapter 12. The adjustment or lack thereof is related to a number of considerations, especially whether the comparisons were planned prior to the analysis.

There are four common approaches in the educational and behavioral sciences when it comes to comparing multiple groups. These are

- Planned comparisons without an omnibus test

- The omnibus test followed by unplanned but unadjusted group comparisons

- The omnibus test followed by unplanned but adjusted group comparisons

- Unplanned but adjusted group comparisons without the omnibus test

Each approach has merits, and each has been criticized from different statistical, philosophical, and practical perspectives. In the remainder of this chapter, the first approach, planned comparisons without an omnibus test, is discussed. Chapter 12 presents the last three methods, as well as highlighting the strengths and weaknesses of all four approaches. By understanding the perspectives surrounding each method, a decision can be made regarding the most useful approach for dealing with the research problem at hand.

11.1 PLANNED COMPARISONS

In planning a research study, researchers typically have in mind a *specific* set of hypotheses that the experiment is designed to test. This may also be the case in either experimental or nonexperimental research. In general, planned comparisons are derived directly from research questions that motivate a researcher to plan a study in the first place. They are so named because the comparisons that the researcher wishes to examine are planned in advance of exploring the data.[1] Planned comparisons, often few in number, are based on the researcher's substantive knowledge and theoretical work in the field.

11.2 EXAMINATION OF WEIGHT LOSS CONDITIONED ON DIET

Consider a 12-month study that compared 4 weight-loss diet regimines representing a spectrum of low to high carbohydrate intake (Gardner et al., 2007). The 240 participants, overweight, nondiabetic, premenopausal women, were randomly assigned to follow the Atkins, Zone, LEARN, or Ornish diets ($n = 60$ in each group). Each participant received weekly instruction for 2 months, then an additional follow-up

---

[1] Planned comparisons are sometimes referred to as a priori comparisons or hypotheses.

at 10 months. After 12 months, the participants' change in weight (in kg) was recorded. The data, contained in *Diet.csv*, will be used to answer the following research questions, which were espoused prior to examining the data.

1. Is there a difference in weight loss between subjects assigned to the most carbohydrate-restrictive diet (Atkins) and subjects assigned to the least carbohydrate-restrictive diet (Ornish)?

2. Is there a difference in weight loss between subjects assigned to the most carbohydrate-restrictive diet (Atkins) and subjects assigned to lesser carbohydrate-restrictive diets (Zone, LEARN, and Ornish)?

3. Is there a difference in weight loss between subjects assigned to a carbohydrate-restrictive diet (Atkins and Zone) and subjects assigned to a behavior modification diet (LEARN and Ornish)?

The analysis of these comparisons begins like any other, by graphically and numerically exploring the data. Command Snippet 11.1 shows the syntax to read in the *Diet.csv* data and examine the data frame object. This will be used to examine the three planned research questions.

Command Snippet 11.1: Syntax to read in the *Diet.csv* data, examine the data frame object, and obtain graphical and numerical summaries for each group.

```
## Read in the data
> diet <- read.table(file = "/Documents/Data/Diet.csv", header
   = TRUE, sep = ",", row.names = "ID")

## Examine the data frame object
## Output is suppressed
> head(diet)
> tail(diet)
> str(diet)
> summary(diet)
```

11.2.1   Exploration of Research Question 1

Command Snippet 11.2 shows the syntax to create the groups and obtain graphical and numerical summaries necessary to examine differences in weight loss between study participants assigned to the Atkins diet and study participants assigned to the Ornish diet. The subset() function, introduced in Chapter 2, is initially used to create a new data frame called diet.1 containing only the data from participants assigned to the Atkins and Ornish diets.

Figure 11.1 shows the density plots for the weight loss for study participants assigned to the Atkins diet and study participants assigned to the Ornish diet. These plots show that there are sample differences in weight loss between subjects that were assigned to these diets. Those assigned to the Atkins diet ($M = -14$) lost

Command Snippet 11.2: Syntax to examine the first research question.

```
## Subset the data frame
> diet1 <- subset(x = diet, subset = Diet == "Atkins" | Diet ==
    "Ornish")

## Density plots of the data
> plot(density(diet1$WeightChange[diet1$Diet == "Ornish"]),
    main = " ", xlab = "12-Month Weight Change", xlim = c(-80,
    40), bty = "l")
> lines(density(diet1$WeightChange[diet1$Diet == "Atkins"]),
    lty = "dashed")
> legend(x = -75, y = 0.025, legend = c("Atkins", "Ornish"),
    lty = c("dashed", "solid"), bty = "n")

## Conditional means
> tapply(X = diet1$WeightChange, INDEX = diet1$Diet, FUN =
    mean)
     Atkins       LEARN      Ornish        Zone
 -14.482533          NA   -5.943527          NA

## Conditional standard deviations
> tapply(X = diet1$WeightChange, INDEX = diet1$Diet, FUN = sd)
   Atkins    LEARN   Ornish     Zone
 14.91540       NA 14.95681       NA
```

more weight, on average, than those assigned to the Ornish diet (-6). However, there was overlap of the distributions, as shown in Figure 11.1. In both diets some of the subjects lost weight, while others gained weight. The Atkins diet seems to have had better success in weight loss with approximately 75% of the subjects assigned to the Atkins diet losing weight, while only about 50% of the subjects assigned to the Ornish diet lost weight. The variation in weight loss is roughly equal for both diets $(SD = 15$ for both diets).

11.2.2 Exploration of Research Question 2

Command Snippet 11.3 shows the syntax to create the groups and graphical and numerical summaries necessary to examine difference in weight loss between study participants assigned to the Atkins diet and study participants assigned to the other three diets. A new data frame called diet.2 is created from the original data. The levels of the Diet factor in this new data frame are then reassigned using the levels() function.

Figure 11.2 shows the density plots for the weight loss for study participants assigned to the Atkins diet and study participants assigned to the other three diets combined. These plots show that there were differences in weight loss between the subjects assigned to Atkins $(M = -14)$ and the subjects assigned to the other three diets $(M = -6)$. Subjects assigned to the Atkins diet typically lost more weight than the subjects assigned to the other diets. The plots also suggest that there is more

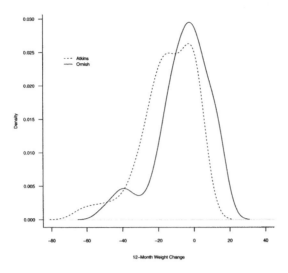

Figure 11.1: Conditional density plots showing the distribution of weight loss for the Atkins and the Ornish diets.

variation in weight loss for subjects on the Atkins diet (SD = 15), indicating that the weight loss for subjects on less carbohydrate-restrictive diets (SD = 14)) is slightly more homogeneous.

11.2.3 Exploration of Research Question 3

Command Snippet 11.4 shows the syntax to create the groups and graphical and numerical summaries necessary to examine difference in weight loss between study participants assigned to the carbohydrate-restrictive diets (Atkins/Zone) and study participants assigned to the behavior modification diets (LEARN/Ornish). A new data frame called diet.3 is created from the original data. The levels of the Diet factor in this new data frame are then reassigned using the levels() function.

Figure 11.3 shows the density plots for the weight loss for study participants assigned to the Atkins or Zone diets and subjects assigned to the LEARN or Ornish diets. These plots highlight differences in weight loss between the subjects assigned to carbohydrate-restrictive diets and the subjects assigned to behavior modification diets, although the differences were minimal. The typical weight loss for participants assigned to a carbohydrate-restrictive diet ($M = -10$) was, on average, higher than for participants assigned to a behavior modification diet ($M = -7$). The variation in weight loss was similar for these groups (SD = 15 and SD = 14, respectively).

Command Snippet 11.3: Syntax to examine the second research question.

```
## Create new data frame
> diet2 <- diet

## Reassign levels of diet
> levels(diet2$Diet) <- c("Atkins", "Others", "Others",
    "Others")

## Density plots of the data
> plot(density(diet2$WeightChange[diet2$Diet == "Others"]),
    main = " ", xlab = "12-Month Weight Change", xlim = c(-80,
    40), bty = "l")
> lines(density(diet2$WeightChange[diet2$Diet == "Atkins"]),
    lty = "dashed")
> legend(x = -75, y = 0.030, legend = c("Atkins", "Others"),
    lty = c("dashed", "solid"), bty = "n")

## Conditional means
> tapply(X = diet2$WeightChange, INDEX = diet2$Diet, FUN =
    mean)
    Atkins      Others
-14.482533   -6.245896

## Conditional standard deviations
>   tapply(X = diet2$WeightChange, INDEX = diet2$Diet, FUN = sd)
  Atkins     Others
14.91540   13.73757
```

11.3 FROM RESEARCH QUESTIONS TO HYPOTHESES

In order to answer the research questions, a test statistic is chosen that will quantify group differences in the sample data. Mean differences, then, will be the basis for examining group differences in the sample, and also the objects of inference. That is, the goal is to make inferences about population mean differences among the groups. This in turn will be used to express the null hypotheses for the comparisons. Using the mean difference, the null hypothesis for each of the three research questions can be delineated as the following.

1. H_{01} : There is no difference in the average weight loss between subjects assigned to the Atkins diet and subjects assigned to the Ornish diet.

2. H_{02} : There is no difference in average weight loss between subjects assigned to the Atkins diet and subjects assigned to the other diets (combined).

3. H_{03} : There is no difference in average weight loss between subjects assigned to the carbohydrate-restrictive diets (Atkins/Zone) and subjects assigned to the behavior modification diets (LEARN/Ornish).

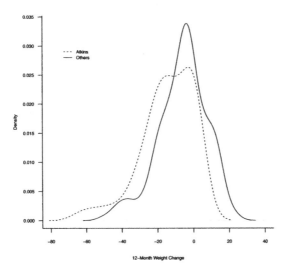

Figure 11.2: Conditional density plots showing the distribution of weight loss for the Atkins and the other three diets.

To test each of these hypotheses, the test statistic computed from the observed data needs to be evaluated as to whether or not it is unlikely given the expected variation across the potential random assignments. Before this can be done, however, the test statistic needs to be computed. With many groups, the statistic is slightly more complicated than just the simple arithmetic means of the groups indicated in the research questions. In the next section, the *contrast* is introduced as a general index of differences among group means.

11.4 STATISTICAL CONTRASTS

A contrast is a statistical comparison of two or more group means. More formally, it is a mathematically expressed sum (or difference) of the group means, where each mean is multiplied by some coefficient (or weight). These expressions, called *linear combinations*, can be symbolically written as

$$\Psi = w_1(\mu_1) + w_2(\mu_2) + w_3(\mu_3) + \cdots + w_k(\mu_k), \qquad (11.1)$$

where k is the number of groups. The weights, $w_1, w_2, w_3, \ldots, w_k$, take either positive or negative values with the caveat that the set of weights must sum to zero.[2]

---

[2]When the groups have unequal sample sizes (unbalanced design), this sum is adjusted using the sample sizes as weights.

Command Snippet 11.4: Syntax to examine the third research question.

```
## Create new data frame
> diet3 <- diet

## Reassign levels of diet
> levels(diet3$Diet) <- c("CR", "BM","BM","CR")

## Density plots of the data
> plot(x = density(x = diet3$WeightChange[diet3$Diet == "BM"]),
    main = " ", xlab = "12-Month Weight Change", xlim = c(-80,
    45), bty = "l")
> lines(density(diet3$WeightChange[diet3$Diet == "CR"]), lty =
    "dashed")
> legend(x = -75, y = 0.030, legend =
    c("carbohydrate-restrictive", "Behavior Modification"), lty
    = c("dashed", "solid"), bty = "n")

## Conditional means
> tapply(X = diet3$WeightChange, INDEX = diet3$Diet, FUN =
    mean)
      CR        BM
-9.873035 -6.737075

## Conditional standard deviations
>  tapply(X = diet3$WeightChange, INDEX = diet3$Diet, FUN = sd)
     CR        BM
14.51608 14.28824
```

In Equation 11.1, μ_j is the population mean for the jth group. The values for the set of weights are chosen by the researcher to reflect the comparisons that he or she wants to test; several examples are considered below.

Contrasts have actually been previously mentioned, but they were not labeled as such for purposes of brevity. A contrast was used every time differences between two groups were tested in Chapters 6, 7, and 9—although they were not explicitly referred to as contrasts in those chapters. For example, in Chapter 7, the null hypothesis that the average educational level for Mexican immigrants was equal to the average educational level for non-Mexican immigrants was tested. This difference in population means can symbolically be expressed as

$$H_0 : \mu_{\text{Mexican}} - \mu_{\text{non-Mexican}} = 0.$$

In this hypothesis, the contrast is the difference in means, or

$$\Psi = \mu_{\text{Mexican}} - \mu_{\text{non-Mexican}},$$

which is equivalent to

$$\Psi = (1)\mu_{\text{Mexican}} + (-1)\mu_{\text{non-Mexican}}.$$

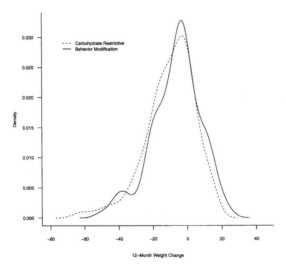

Figure 11.3: Conditional density plots showing the distribution of weight loss for the Atkins/Zone and the LEARN/Ornish diets.

In this contrast, the weight for the first group (Mexicans) is positive one, and the weight for the second group (non-Mexicans) is negative one. The sum of these two weights is zero. The opposite signs on the contrast weights indicate the means which are being compared. A contrast created for testing the differences between two group means is called a *simple* or *pairwise* contrast. The contrast is then written as a null hypothesis,

$$H_0 : \Psi = 0. \tag{11.2}$$

Notice that the contrast is not equivalent to the null hypothesis, but represents only the differences among the mean within the null hypothesis. The simple contrast regarding the difference in average weight loss between subjects assigned to the most carbohydrate-restrictive diet (Atkins) and subjects assigned to the least carbohydrate-restrictive diet (Ornish) can be expressed as

$$\Psi_1 = (1)\mu_{\text{Atkins}} + (0)\mu_{\text{LEARN}} + (-1)\mu_{\text{Ornish}} + (0)\mu_{\text{Zone}},$$

where the subscript 1 in Ψ_1 indicates that this is the first contrast. Notice that all the group means are considered, but the means for the LEARN and Zone diets drop out of the contrast. Even though the interest is in only two of the groups, the contrast is written using all of the groups. By using a weight of zero for both the Zone and LEARN means, the comparison will not include these two groups. The null hypothesis, again denoted with a subscript of 1, is

$$H_{01} : (1)\mu_{\text{Atkins}} + (0)\mu_{\text{LEARN}} + (-1)\mu_{\text{Ornish}} + (0)\mu_{\text{Zone}} = 0,$$

or more compactly,

$$H_{01} : \mu_{\text{Atkins}} - \mu_{\text{Ornish}} = 0.$$

11.4.1 Complex Contrasts

Contrasts can also be formulated to make comparisons between more than two groups. These are called *complex* contrasts. For example, the second and third research questions listed in Section 11.2 involve testing complex contrasts. In the second research question the goal is to test whether there are differences in the average weight loss between subjects assigned to the most carbohydrate-restrictive diet (Atkins) and subjects assigned to lesser carbohydrate-restrictive diets (Zone, LEARN, and Ornish combined). This contrast can be expressed as

$$\Psi_2 = \mu_{\text{Atkins}} - \frac{\mu_{\text{LEARN}} + \mu_{\text{Ornish}} + \mu_{\text{Zone}}}{3}.$$

The entire contrast can be multiplied by 3 to clear the fraction so that

$$(3)\Psi_2 = (3)\mu_{\text{Atkins}} - (\mu_{\text{LEARN}} + \mu_{\text{Ornish}} + \mu_{\text{Zone}}).$$

Each of the means for LEARN, Ornish and Zone can be multiplied by -1 to yield

$$(3)\Psi_2 = (3)\mu_{\text{Atkins}} + (-1)\mu_{\text{LEARN}} + (-1)\mu_{\text{Ornish}} + (-1)\mu_{\text{Zone}}.$$

Under the null hypothesis, $(3)\Psi_2 = 0$, and thus, $\Psi_2 = 0$. Therefore, the multiplier of Ψ_2 can be disregarded and the null hypothesis can be expressed as,

$$H_{02} : (3)\mu_{\text{Atkins}} + (-1)\mu_{\text{LEARN}} + (-1)\mu_{\text{Ornish}} + (-1)\mu_{\text{Zone}} = 0.$$

This is a complex contrast as more than two means have weights that are not zero (all the weights are nonzero in this case). As usual, the weights sum to zero. The third research question requires us to examine whether there are differences in the average weight loss between subjects assigned to a carbohydrate-restrictive diet (Atkins and Zone) and subjects assigned to a behavior modification diet (LEARN and Ornish). Symbolically this contrast can be expressed as

$$\Psi_3 = \frac{\mu_{\text{Atkins}} + \mu_{\text{Zone}}}{2} - \frac{\mu_{\text{LEARN}} + \mu_{\text{Ornish}}}{2}.$$

Following the same process to remove the fractions, carrying through the negative, and adding contrast weights, the linear contrast is

$$(2)\Psi_3 = (1)\mu_{\text{Atkins}} + (-1)\mu_{\text{LEARN}} + (-1)\mu_{\text{Ornish}} + (1)\mu_{\text{Zone}},$$

and the null hypothesis for the contrast as

$$H_{03} : (1)\mu_{\text{Atkins}} + (-1)\mu_{\text{LEARN}} + (-1)\mu_{\text{Ornish}} + (1)\mu_{\text{Zone}} = 0,$$

or collecting terms,

$$H_{03} : (\mu_{\text{Atkins}} + \mu_{\text{Zone}}) - (\mu_{\text{LEARN}} + \mu_{\text{Ornish}}).$$

In each of the contrasts, whether they are simple or complex, opposite signs on the contrast weights indicate which groups are being compared. For example, in the third contrast, both the Atkins and Zone means have a positive contrast weight, whereas the means for LEARN and Ornish have a negative contrast weight. This indicates that the comparison is between the Atkins and Zone diets combined and the LEARN and Ornish diets combined.

11.5 COMPUTING THE ESTIMATED CONTRASTS USING THE OBSERVED DATA

For each of the three contrasts, a point estimate is computed using the observed data. This estimate is denoted as $\hat{\Psi}$, and is computed by substituting the observed sample means for the population means in the contrast of interest. Each of the three contrasts estimates for the diet comparisons are listed and computed below.

For the first comparison, recall that the contrast denoted Ψ_1 is

$$\Psi_1 = (1)\mu_{\text{Atkins}} + (0)\mu_{\text{LEARN}} + (-1)\mu_{\text{Ornish}} + (0)\mu_{\text{Zone}}.$$

Using the sample mean weight loss for each of the diets from the exploratory analysis, the contrast can be estimated as

$$\begin{aligned} \hat{\Psi}_1 &= (1)\mu_{\text{Atkins}} + (0)\mu_{\text{LEARN}} + (-1)\mu_{\text{Ornish}} + (0)\mu_{\text{Zone}} \\ &= (1)(-14.5) + (0)(-7.5) + (-1)(-5.9) + (0)(-5.3) \\ &= -14.5 + 5.9 \\ &= -8.6. \end{aligned}$$

To reiterate, the weights are set by the researcher and the means are estimated based on the sample data. The estimated difference using the observed data in average weight loss between subjects assigned to the most carbohydrate-restrictive diet and subjects assigned to the least carbohydrate-restrictive diet is 8.6 kg. This estimate can also be directly computed using R. To carry out the computation a vector of the contrast weights is created. To determine the order of the contrast weights, say $w_j = [1, 0, -1, 0]$ vs. $w_j = [1, 0, 0, -1]$, the order that the means are printed when the tapply() function is executed can be examined. This is typically in alphabetical order by level name. Recall that the levels() function is used to obtain the levels of the factor of interest. The vector of contrast weights is then multiplied by the vector of sample group means obtained through the tapply() function. Lastly,

Command Snippet 11.5: Syntax to compute the estimate of the first contrast.

```
## Determine the order of the levels
> levels(diet$Diet)
[1] "Atkins" "LEARN"  "Ornish" "Zone"

## Create a vector of the contrast weights
> con1 <- c(1, 0, -1, 0)

## Multiply the contrast weights and sample means together
> tapply(X = diet$WeightChange, INDEX = diet$Diet, FUN = mean)
    * con1
    Atkins       LEARN     Ornish        Zone
-14.482533    0.000000   5.943527    0.000000

## Sum the products
> sum(tapply(X = diet$WeightChange, INDEX = diet$Diet, FUN =
    mean) * con1)
[1] -8.539006
```

these products are added together using the sum() function. Command Snippet 11.5 shows the syntax to carry out the computation for estimating the first contrast ($\hat{\Psi}_1$).

Similarly, the estimates for the other two complex contrasts are computed. Command Snippet 11.6 shows the syntax for finding these sample estimates. The sample data suggest that the difference in the average weight loss between subjects assigned to the most carbohydrate-restrictive diet and subjects assigned to lesser carbohydrate-restrictive diets is 24.7 kg. These data also suggest that there is, on average, a difference in the weight loss between subjects assigned to a carbohydrate-restrictive diet and subjects assigned to a behavior modification diet of 6.3 kg.

Command Snippet 11.6: Syntax to compute the estimates for the two complex contrasts.

```
## Create the vectors of the contrast weights
> con2 <- c(3,-1,-1,-1)
> con3 <- c(1,-1,-1,1)

## Compute the estimate for the second contrast
> sum(tapply(X = diet$WeightChange, INDEX = diet$Diet, FUN =
    mean) * con2)
[1] -24.70991

## Compute the estimate for the third contrast
> sum(tapply(X = diet$WeightChange, INDEX = diet$Diet, FUN =
    mean) * con3)
[1] -6.27192
```

11.6 TESTING CONTRASTS: RANDOMIZATION TEST

The question of whether each of these sample differences is "real" or just due to the variability expected under random assignment is now addressed. After computing the observed estimate for each contrast, each of the null hypotheses can be tested. Since the study participants were randomly assigned to diets, the randomization test can be used to examine the variation expected in each of these contrasts because of random assignment. Figure 11.4 presents the steps used to carry out a randomization test to obtain the distribution for a particular contrast under the null model of no difference.

Randomization Test for a Contrast

- Randomly permute the observed sample data.

- Compute $\hat{\Psi}$—the estimated contrast from the permuted data.

- Repeat these first two steps many times, say R times, each time recording the estimated contrast $\hat{\Psi}$.

- The distribution of $\hat{\Psi}_1, \hat{\Psi}_2, \hat{\Psi}_3, \ldots, \hat{\Psi}_R$ can be used as an estimate of the sampling distribution of Ψ under the null model. The value of $\hat{\Psi}$ from the observed data can be evaluated using this distribution.

Figure 11.4: Steps used to carry out a randomization test to obtain the distribution for a particular contrast under the null model of no difference.

The only difference between the above randomization test and the one introduced in Chapter 6 for the simple mean difference is that with complex contrasts, the weighted mean difference (the estimated value of the contrast) is being computed for each permutation of the data rather than the arithmetic mean difference. Otherwise, the method is completely the same. Command Snippet 11.7 shows the syntax for permuting the observed data 4999 times, a function to compute the contrast value for the first contrast, the application of that function to the permuted data, and the calculation of the Monte Carlo p-value. The output in Command Snippet 11.8 shows that the p-value is extremely small for the first contrast. This is strong evidence that the difference between the Ornish and Atkins diet is a "real" population difference.

The randomization test for the other two contrasts is carried out in a similar manner. The permuted samples that have already been collected for the first test can be used in each of the other two contrast tests. Command Snippet 11.8 and Command Snippet 11.9 show the syntax for carrying out the randomization test for the second and third contrasts, respectively. Based on these results, there is overwhelming evidence that there are mean differences between participants assigned to the Atkins diet and those assigned to the Ornish diet ($p = 0.001$). There is also overwhelming evidence that there are mean differences between participants assigned to the Atkins diet and participants assigned to the other three diets ($p = 0.0002$). There is, however, only borderline evidence to support that there are mean differences in the weight loss between participants assigned to carbohydrate-restrictive diets (Atkins/Zone) and participants assigned to behavior modification diets (LEARN/Ornish; $p = 0.10$).

Command Snippet 11.7: Set of commands to carry out the randomization test for the first contrast.

```
## Permute the WeightChange scores
> permuted <- replicate(n = 4999, expr =
    sample(diet$WeightChange))

## Function to compute the estimated contrast
> contrast.1 <- function(data) {
  (1) * mean(data[1:60]) +
  (0) * mean(data[61:120]) +
  (-1) * mean(data[121:180]) +
  (0) * mean(data[181:240])
  }

## Check the function with the observed data
> contrast.1(diet$WeightChange)
[1] -8.539006

## Apply the contrast to each of the 4999 permuted samples
> perm.contrasts.1 <- apply(X = permuted, MARGIN = 2, FUN =
    contrast.1)

## Examine the distribution of the contrast for the
##permuted samples
> plot(density(perm.contrasts.1))
> mean(perm.contrasts.1)
[1] 0.005443845
> sd(perm.contrasts.1)
[1] 2.629473

## Calculate the Monte Carlo p-value
> length(diffs[abs(perm.contrasts.1) >= 8.6])
[1] 4
> (4 + 1) / (4999 + 1)
[1] 0.0009998
```

11.7 STRENGTH OF ASSOCIATION: A MEASURE OF EFFECT

The results of any statistical test should always be accompanied by a measure of effect regardless the degree of evidence proffered by the *p*-value. A conventional supplement to the results of a contrast test is to provide a measure of the *strength*

Command Snippet 11.8: Set of commands to carry out the randomization test for the second contrast.

```
## Function to compute the estimated contrast
> contrast.2 <- function(data) {
  (3) * mean(data[1:60]) +
  (-1) * mean(data[61:120]) +
  (-1) * mean(data[121:180]) +
  (-1) * mean(data[181:240])
  }

## Check the function with the observed data
> contrast.2(diet$WeightChange)
[1] -8.539006

## Apply the contrast to each of the 4999 permuted samples
> perm.contrasts.2 <- apply(X = permuted, MARGIN = 2, FUN =
    contrast.2)

## Examine the distribution of the contrast for the
##permuted samples
> plot(density(perm.contrasts.2))
> mean(perm.contrasts.2)
[1] 0.03796324
> sd(perm.contrasts.2)
[1] 6.5019

## Calculate the Monte Carlo p-value
> length(perm.contrasts.2[abs(perm.contrasts.2) >= 24.7])
[1] 0
> (0 + 1) / (4999 + 1)
[1] 0.00019996
```

of the association between the group differences identified in the contrast and the outcome variable. The most popular measure of association is eta-squared (η^2)—also called the correlation ratio (Huberty, 2002).

Eta-squared provides an estimate of the variation in the outcome variable accounted for by differences in contrast groups, and is computed separately for each of the a priori linear contrasts. Eta-squared is computed as

$$\eta^2 = \frac{SS_{\Psi_k}}{SS_{Total}}, \tag{11.3}$$

where SS_{Ψ_k} is the sum of squares for the kth contrast, and SS_{Total} is the total sum of squares.

11.7.1 Total Sum of Squares

The total sum of squares, SS_{Total}, is a measure of the total variation in the observed outcome variable. In order to quantify this variation, the squared deviations from the

Command Snippet 11.9: Set of commands to carry out the randomization test for the third contrast.

```
## Function to compute the estimated contrast
> contrast.3 <- function(data) {
  (1) * mean(data[1:60]) +
  (-1) * mean(data[61:120]) +
  (-1) * mean(data[121:180]) +
  (1) * mean(data[181:240])
  }

## Check the function with the observed data
> contrast.3(diet$WeightChange)
[1] -6.27192

## Apply the contrast to each of the 4999 permuted samples
> perm.contrasts.3 <- apply(X = permuted, MARGIN = 2, FUN =
    contrast.3)

## Examine the distribution of the contrast for the
##permuted samples
> plot(density(perm.contrasts.3))
> mean(perm.contrasts.3)
[1] -0.06031592
> sd(perm.contrasts.3)
[1] 3.732738

## Calculate the Monte Carlo p-value
> length(perm.contrasts.3[abs(perm.contrasts.3) >= 6.3])
[1] 477
> (477 + 1) / (4999 + 1)
[1] 0.0956
```

mean for each of the observed values of the outcome variable is summed. In general the computation is

$$SS_{Total} = \sum (Y_i - \bar{Y})^2, \qquad (11.4)$$

where Y_i is the ith observation and $\bar{Y}$ is the marginal mean (sometimes referred to as the *grand mean*). To compute the total sum of squares for the weight change variable, the sum of squared deviations from the marginal mean of -8.3 is computed for each of the 240 observed weight change values as

$$SS_{Total} = (-63.4 + 8.3)^2$$
$$+ (-55.6 + 8.3)^2 + (-48.5 + 8.3)^2 + \cdots + (15.1 + 8.3)^2$$
$$= 49959.$$

This value is a quantification of the total variation in the observed data. This is easily computed from the sample variance, which is defined as

$$s_Y^2 = \frac{\sum (Y_i - \bar{Y})^2}{n - 1}$$

$$= \frac{SS_{Total}}{N - 1},$$

(11.5)

where N is the total sample size, $N = n_1 + n_2 + \cdots + n_k$. Thus, the total sum of squares can be computed by multiplying the sample variance by the quantity $N - 1$. Command Snippet 11.10 shows this computation for the observed weight changes.

Command Snippet 11.10: Computation of the total sum of squares for the observed weight changes.

```
> var(diet$WeightChange) * (length(diet$WeightChange) - 1)
[1] 49959.62
```

11.8 CONTRAST SUM OF SQUARES

The contrast sum of squares, $SS_{\hat{\Psi}_k}$, is a measure of the variation in the observed outcome variable that can be explained via the groups identified in the contrast. This can be found using

$$SS_{\hat{\Psi}_k} = \frac{\hat{\Psi}_k^2}{\sum \frac{w_j^2}{n_j}},$$

(11.6)

where $\hat{\Psi}_k$ is the estimated value of the contrast, w_j is the contrast weight associated with the jth group, and n_j is the number of observed values (i.e., sample size) for the jth group. To calculate the contrast sum of squares for the first contrast, the following is used:

$$SS_{\hat{\Psi}_1} = \frac{(-8.5)^2}{\frac{(1)^2}{60} + \frac{(0)^2}{60} + \frac{(-1)^2}{60} + \frac{(0)^2}{60}}$$

$$= \frac{72.9}{0.03}$$

$$= 2187.$$

(11.7)

This can also be computed using the syntax in Command Snippet 11.11.

11.9 ETA-SQUARED FOR CONTRASTS

The effect size for the contrast, eta-squared $\eta_{\hat{\Psi}_k}^2$, is the ratio of the contrast sum of squares and the total sum of squares, namely,

$$\eta^2_{\hat{\Psi}_k} = \frac{SS_{\hat{\Psi}_k}}{SS_{\text{Total}}}. \tag{11.8}$$

Using the values computed thus far, the effect size for the first contrast is

$$\eta^2_{\hat{\Psi}_1} = \frac{2187}{49,960}$$
$$= 0.044.$$

This implies that 4.4% of the variation in weight change can be accounted for by differences between the Atkins and Ornish diets. The eta-squared value for the other two contrasts can be computed in a similar manner. Command Snippet 11.12 shows the syntax for computing the contrast sum of squares and the eta-squared value for the second and third contrasts. The total sum of squares does not change from contrast to contrast so that value is not recomputed.

Command Snippet 11.11: Computation of the contrast sum of squares for the first contrast.

```
## Numerator
> psi <- sum(tapply(X = diet$WeightChange, INDEX = diet$Diet,
    FUN = mean) * con1)
> numerator <- psi ^ 2
> numerator
[1] 72.91463

## Denominator
> denominator <- sum(con1 ^ 2 / table(diet$Diet))
> denominator
[1] 0.03333333

## Contrast sum of squares
> numerator / denominator
[1] 2187.439
```

11.10   BOOTSTRAP INTERVAL FOR ETA-SQUARED

Just as was done for the distance measures of effect introduced in Chapter 9, an interval estimate for each of the eta-squared estimates should also be computed and reported. The confidence interval for these measures can also be bootstrapped. The function used in the boot() function will need to compute the estimate of eta-squared for each bootstrap replicate rather than an estimate of Cohen's d. In Command Snippet 11.13 syntax is provided to bootstrap the interval estimate for the effect size for the first contrast using a nonparametric bootstrap. The computation for the other two effect size estimates is left as an exercise for the reader.

11.11   SUMMARIZING THE RESULTS OF A PLANNED CONTRAST TEST ANALYSIS

An example write-up reporting the results for the first planned contrast in the diet example is given below. The write-ups for the other two planned contrasts are again left as an exercise for the reader.

Sample Write-Up

Twelve-month weight change in overweight, nondiabetic, premenopausal women was analyzed using a series of linear contrasts to examine three a priori hypotheses about differences between diets. A randomization test was used to determine whether each of the contrasts was statistically reliable. A Monte Carlo corrected p-value (Davison & Hinkley, 1997) for each contrast was computed by permuting the data 4999 times.

The first a priori hypothesis to be examined was whether or not there were differences in weight change between study participants assigned to the most carbohydrate-restrictive diet (Atkins) and study participants assigned to the least carbohydrate-restrictive diet (Ornish). The test of the contrast suggests that these data provide overwhelming evidence against the null hypothesis of no difference ($\hat{\Psi} = -8.5, p = 0.001$). The sample data showed that the subjects assigned to the Atkins diet ($M = -14.5$ kg) lost more weight, on average, than subjects assigned to the Ornish diet ($M = -5.9$ kg). These differences in diet account for an estimated 4.4% of the variation in weight change as measured by eta-squared. A 95% bias-corrected-and-accelerated nonparametric bootstrap interval using 4999 replications was also computed for this effect as $[0.005, 0.107]$.

11.12   EXTENSION: ORTHOGONAL CONTRASTS

Educational and behavioral researchers in some circumstances choose their contrast weights to produce *orthogonal contrasts*. Orthogonal contrasts are contrasts designed to test completely uncorrelated hypotheses. This means that when a contrast is tested it does not provide any information about the results of subsequent contrasts that are orthogonal to it (Kirk, 1995). One advantage of orthogonal contrasts is that their independent influence on the total effect size can be determined.

Suppose there are two linear contrasts, the first having weights a_i and the second having weights b_i. The two contrasts are orthogonal if

$$\sum_{i=1}^{j} a_i b_i = a_1 b_1 + a_2 b_2 + a_3 b_3 + \cdots + a_j b_j = 0, \tag{11.9}$$

Command Snippet 11.12: Computation of the contrast sum of squares and eta-squared for the second and third contrasts.

```
## Contrast 2
> numerator <- sum(tapply(X = diet$WeightChange, INDEX =
    diet$Diet, FUN = mean) * con2) ^ 2
> denominator <- sum(con2 ^ 2 / table(diet$Diet))

## Contrast SS
> numerator / denominator
[1] 18317.39

## Eta-squared estimate
> 18317 / 49960
[1] 0.3666333

## Contrast 3
> numerator <- sum(tapply(X = diet$WeightChange, INDEX =
    diet$Diet, FUN = mean) * con3) ^ 2
> denominator <- sum(con3 ^ 2 / table(diet$Diet))

## Contrast SS
> numerator / denominator
[1] 590.0546

## Eta-squared estimate
> 590 / 49960
[1] 0.01180945
```

assuming that the group sample sizes are equal. As an example, consider the sets of contrast weights for each of the three a priori contrasts introduced earlier in this chapter:

$$w_{\Psi_1} = [1, 0, -1, 0],$$
$$w_{\Psi_2} = [3, -1, -1, -1],$$
$$w_{\Psi_3} = [1, -1, -1, 1].$$

The first and second contrasts are not orthogonal since

$$(1)(3) + (0)(-1) + (-1)(-1) + (0)(-1) = 4 \neq 0.$$

R can be used to examine whether the two contrasts are orthogonal by multiplying the vectors of contrast weights together and then using the sum() function on the product to see if the result is zero. Command Snippet 11.14 shows the syntax for the computations to examine the orthogonality of the first and third contrasts, as well as the second and third contrasts.

In a family of several contrasts, if every pair of contrast weights shows the property that the sum of the corresponding products is zero (i.e., the contrasts are orthogonal), the set or family of contrasts is considered to be *mutually orthogonal*. A researcher

Command Snippet 11.13: Use of the nonparametric bootstrap to compute an interval estimate for eta-squared for the first contrast.

```
## Function to compute eta-squared
> eta.squared <- function(data, indices) {
  d <- data[indices,]
  num <- sum(tapply(X = d$WeightChange, INDEX = d$Diet, FUN =
      mean) * con1) ^ 2
  den <- sum(con1 ^ 2 / table(d$Diet))
  SS.con <- num / den
  SS.tot <- var(d$WeightChange) * 239
  SS.con / SS.tot
  }

## Test eta.squared() function
> eta.squared(diet)
[1] 0.04378414

## Nonparametric bootstrap
> library(boot)
> nonpar.boot <- boot(data = diet, statistic = eta.squared, R =
      4999, strata = diet$Diet)

## Examine the bootstrap distribution
> plot(density(nonpar.boot$t), xlab = "Bootstrapped
    Eta-Squared", main = " ")
> nonpar.boot

STRATIFIED BOOTSTRAP


Call:
boot(data = diet, statistic = eta.squared, R = 4999, strata =
    diet$Diet)


Bootstrap Statistics :
      original        bias     std. error
t1* 0.04378414  0.004400037   0.02721602

## Bias-corrected-and-accelerated interval
> boot.ci(boot.out = nonpar.boot, type = "bca")
BOOTSTRAP CONFIDENCE INTERVAL CALCULATIONS
Based on 4999 bootstrap replicates

CALL :
boot.ci(boot.out = nonpar.boot, type = "bca")

Intervals :
Level        BCa
95%   ( 0.0051,  0.1071 )
Calculations and Intervals on Original Scale
```

must examine *each pair* of contrast's weights when determining if a set of contrasts is mutually orthogonal. If there are g groups, one can find a set of $g - 1$ contrasts that are mutually orthogonal.

That is, the total effect size for comparing the g groups can be partitioned exactly into the set of $g - 1$ mutually orthogonal contrasts. It is important to note that when contrasts are not orthogonal, the independent contributions of the contrasts to the total effect cannot be determined. Each of the individual contrasts is tainted by redundant information and misleading to an unknown extent. Redundancy also is caused by the design not being balanced, meaning the group sample sizes are not equal.

Command Snippet 11.14: Computation to check orthogonality between the three contrasts.

```
## Contrast 1 and contrast 2 are not orthogonal
> sum(con1 * con2)
[1] 4

## Contrast 1 and contrast 3 are not orthogonal
> sum(con1 * con3)
[1] 2

## Contrast 2 and contrast 2 are not orthogonal
> sum(con2 * con3)
[1] 4
```

In spite of the above point, it is recommended that researchers not attempt to construct orthogonal contrasts for the sake of producing uncorrelated hypotheses. Rather, researchers should create contrasts that reflect the important hypotheses that are driving the analysis.

In the context of planned comparisons, one situation where orthogonal contrasts are theorized in advance is conducting a trend analysis using *orthogonal polynomial contrasts*. The analyses discussed thus far have been concerned with identifying differences among group means, where the comparisons represent complex contrasts among groups or simple pairwise comparisons. Polynomial contrasts, on the other hand, are used to examine and test the pattern (or trend) in the group means. This only makes sense when the groups have some type of natural ordering, as when the groups represent increasing doses of a medication. In this case, if the pattern in the plotted group means was, say, U-shaped, the researcher might test whether a quadratic trend is present or whether the bend is just due to random variation. Wikipedia—as well as most algebra textbooks—show plots of several polynomial functions that could be examined. Remember, however, that polynomial contrasts are only useful if the factor separating the groups is at the interval level of measurement. Polynomial contrasts would not be appropriate for the diet data since the factor is categorical in nature and there is no natural ordering of the groups.

To provide additional context, consider the situation in which a researcher hypothesizes that study time has a positive effect on word recall. In other words, the researcher believes that on average, people that study a list of words for a longer period of time will be able to recall more words. Furthermore, the researcher believes that the effects on word recall increase at a quadratic rate. A quadratic effect would imply that the increase is not constant but rather U-shaped—the effect of time spent studying on word recall is smaller for lesser amounts of study time and larger at greater amounts of study time.

To examine this issue, 48 middle school students are randomly assigned to three study time conditions (30 min, 60 min, 90 min). The recall period lasts for 2 min and begins immediately after the study period ends. The hypothetical data for this example can be obtained in the *WordRecall.csv* data set. Command Snippet 11.15 shows the syntax for reading in these data and computing the mean word recall across the three study time conditions.

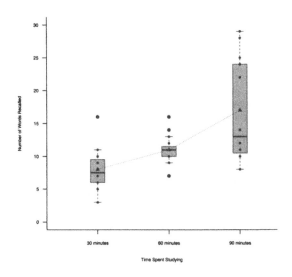

Figure 11.5: Side-by-side box-and-whiskers plots showing the distribution of word recall conditioned on time spent studying. The conditional means are each marked with a triangle. The dotted line, which connects the conditional means, shows a potential polynomial trend.

Figure 11.5 depicts the side-by-side box-and-whiskers plots showing the distribution of word recall conditioned on time spent studying. The conditional means are each marked with a triangle. The dotted line, which connects the conditional means, shows an increasing trend in the average word recall in the sample. Recall that a statistical test can be used to determine whether this trend in the sample means is "real" or whether it is due only to the expected variation in the random assignment.

To test for trends in these means, researchers use a set of contrast weights to produce a set of orthogonal polynomial contrasts. Recall, that there are exactly

$g - 1$ orthogonal polynomial trends that can be tested for g groups. For the word recall example, with three means, two polynomial trends can be tested: (1) a linear trend (straight line), and (2) a quadratic trend (bent line). To determine the contrast weights to test polynomial trends, the `contr.poly()` function in R can be used. This function requires the argument n=, which indicates the number of levels in the factor.

Command Snippet 11.15: Syntax to read in and examine the study data, compute the mean word recall conditional on study time, and plot the conditional distributions.

```
## Read in the Study.csv data
> word <- read.table(file = "/Documents/Data/WordRecall.csv",
    header = TRUE, sep = ",", row.names = "ID")

## Examine the data frame object
## Output is suppressed
> head(word)
> tail(word)
> str(word)
> summary(word)

## Scatterplot of the relationship between recall and study
>  plot(x = word$Study, y = word$Recall, xlab = "Time Spent
     Studying", ylab = "Number of Words Recalled", xlim=c(0.4,
     3.4), ylim = c(0, 30), bty = "l", pch = 20)

## Superimpose the conditional boxplots on the scatterplot
> boxplot(word$Recall[word$Study == 1], word$Recall[word$Study
    == 2], word$Recall[word$Study == 3], at = c(1:3), add =
    TRUE, axes = FALSE, boxwex = 0.2, col = rgb(red = 0.2,
    green = 0.2, blue = 0.2, alpha = 0.3))

## Compute the conditional means
>  tapply(X = word$Recall, INDEX = word$Study, FUN = mean)
 1  2  3
 8 11 17

## Compute the conditional standard deviations
>  tapply(X = word$Recall, INDEX = word$Study, FUN = sd)
       1        2        3
3.011091 2.065591 7.465476

## Compute the conditional sample sizes
> table(word$Study)

 1  2  3
16 16 16
```

Command Snippet 11.16 shows the syntax to find the contrast weights to examine mean trends (linear and quadratic) for three groups. These contrast weights are printed in the columns of the resulting matrix. Thus, to test a quadratic trend, the weights $[0.4082483, -0.8164966, 0.4082483]$ are used. It should be mentioned that the polynomial weights are not easily interpretable and researchers usually rely on computer programs such as R for their derivation.

Command Snippet 11.16: Syntax to find the contrast weights for testing polynomial trends in three group means. The syntax is also provided to assign the trend weights to two separate vectors.

```
## Obtain the polynomial contrast weights
> poly.cont <- contr.poly(3)
> poly.cont
                    .L           .Q
[1,]  -7.071068e-01   0.4082483
[2,]   4.350720e-18  -0.8164966
[3,]   7.071068e-01   0.4082483

## Weights for the linear trend
> con.linear <- poly.cont[ ,1]
> con.linear
[1] -7.071068e-01   4.350720e-18   7.071068e-01

# Weights for the quadratic trend
> con.quad <- poly.cont[ ,2]
> con.quad
[1]   0.4082483  -0.8164966   0.4082483
```

Tests and effect sizes for these contrasts are then carried out in the same manner as any other contrast. It is typical to examine all of the trends and report which ones were found. The computation of these polynomial contrasts and corresponding effect size measures are included as an exercise at the end of this chapter.

11.13 FURTHER READING

The method of contrast analysis has been known since Fisher introduced analysis of variance in the 1930s [Abelson (1962) as cited in Rosenthal and Rosnow (1985)]. A more complete treatment of analysis of contrasts, including analysis for more complex research designs, is provided in Rosenthal, Rosnow, and Rubin (2000). Kirk (1995) and Oehlert (2000) describe contrast analysis among groups with unequal sample sizes. The choice of contrast weights (e.g., orthogonal, trend, etc.) is addressed in Rosenthal et al. (2000), Kirk (1995), and Oehlert (2000). Furthermore, Lomax (2007) and Pedhazur (1997) present the computation of orthogonal contrast weights for groups in which the sample sizes are unequal.

PROBLEMS

11.1 Word recall depends on a number of factors including level of processing and word encoding strategies. Study time can also have an effect on word recall. The data in *WordRecall.csv* contain the number of words recalled for 48 subjects who were randomly assigned to 3 study time conditions (1 = 30 min, 2 = 60 min, and 3 = 90 min).

    a) Test both a linear and quadratic trend in words recalled among the three study conditions using a randomization test.

    b) Compute a point estimate of an effects size for each polynomial contrast and compute the 95% bootstrap interval estimate for each contrast using the BC_a method.

    c) Demonstrate that the eta-squared effect sizes for the linear and quadratic trend add up to the overall between study condition effect size.

11.2   A researcher comes to you for advice about designing an experimental study to compare multiple groups. The researcher is worried about sample size equality across the treatment conditions under investigation. Provide a short report (no more than one page) in which you outline, compare, and contrast the advantages and disadvantages of carrying out the study with both equal sample sizes and unequal sample sizes. Based on what you find, make a recommendation to the researcher about how she should proceed. Reference at least three sources—other than this monograph—that reinforce the rationale of your recommendations.

CHAPTER 12

UNPLANNED CONTRASTS

No aphorism is more frequently repeated in connection with field trials, than that we must ask Nature few questions, or, ideally, one question at a time. The writer is convinced that this view is wholly mistaken. Nature, he suggests, will best respond to a logical and carefully thought out questionnaire; indeed, if we ask her a single question, she will often refuse to answer until some other topic has been discussed.

—R. A. Fisher (1926)

In Chapter 11, methods of inference regarding group differences with more than two groups were examined. The discussion focused on the case in which specific research questions were purposed in advance of collecting the data, and certainly before commencing with any data analysis. These a priori hypotheses were referred to more broadly as planned comparisons. Recall that a priori hypotheses are often few in number, are theoretically grounded, and are scientifically defensible. In this chapter, methods of inference for group differences that fall into three general categories are examined: (1) unplanned, but adjusted group comparisons without the omnibus test; (2) the omnibus test followed by unadjusted group comparisons; and (3) the omnibus test followed by adjusted group comparisons. Each approach has merits, and each has its drawbacks. Even though methodological research for some of these approaches dates back to the 1930s, no resolution as to the single best approach has emerged. In fact, a consensus may never be reached because each approach takes a different statistical, philosophical, and practical perspective to the problem of multiple group comparisons. At the end of the chapter, strengths and weaknesses of each approach are presented. By understanding the perspectives surrounding each method, researchers can decide which approach is most useful for the problem at hand.

To motivate each of the three approaches, the *Diet.csv* data set is once again used. Recall, these data were obtained from a 12-month study that compared four weight-loss regimens on a sample of $N = 240$ overweight, nondiabetic, premenopausal women (Gardner et al., 2007). The 240 participants were randomly assigned to follow the Atkins, Zone, LEARN, or Ornish diets ($n_j = 60$ in each group). Each participant received weekly instruction for 2 months, then an additional 10-month follow-up. Weight loss, in kilograms, at 12 months was the primary outcome. In all three comparison approaches, the goal is to answer the following research questions:

1. Are there mean differences in weight loss for the four diets (Atkins, LEARN, Ornish, and Zone)?

2. If so, how do the diets differ?

12.1 UNPLANNED COMPARISONS

The three approaches for examining group differences between many groups presented in this chapter are exploratory in nature. Unlike the example presented in the Chapter 11, in which group comparisons were planned in advance of examining the data, post-hoc or a posteriori group comparisons are formulated by researchers who do not have specific hypotheses in mind prior to examining their data. As pointed out earlier, educational and behavioral researchers typically take one of three approaches to the exploratory testing of unplanned comparisons:

- The omnibus test followed by unadjusted group comparisons

- The omnibus test followed by adjusted group comparisons

- Unplanned, but adjusted group comparisons without the omnibus test

In the remainder of this chapter, these three methods are presented. At the end of this chapter, the strengths and criticisms of these methods, and that of planned comparisons without the omnibus test, will be discussed.

12.2 EXAMINATION OF WEIGHT LOSS CONDITIONED ON DIET

Consider the comparisons of the weight loss groups from the perspective of no planned hypotheses. This is a scenario in which the researcher has little or no idea of the effectiveness of the diets. The analysis begins like any other, by graphically and numerically exploring the data. Command Snippet 12.1 shows the syntax to read in the *Diet.csv* data and examine the data frame object. The distribution of 12-month weight change is also examined conditioning on diet. Figure 12.1 shows the side-by-side box-and-whiskers plots.

Command Snippet 12.1: Syntax to read in the *Diet.csv* data, examine the data frame object, and obtain graphical and numerical summaries for each group.

```
## Read in the data
> diet <- read.table(file = "/Documents/Data/Diet.csv", header
    = TRUE, sep = ",", row.names = "ID")

## Examine the data frame object
## The output is suppressed.
> head(diet)
> tail(diet)
> str(diet)
> summary(diet)

## Examine the marginal distribution
## The plot is not shown.
> plot(density(diet$WeightChange)

> mean(diet$WeightChange)
[1] -8.305055

> sd(diet$WeightChange)
[1] 14.45808

## Examine the conditional distributions
> boxplot(WeightChange ~ Diet, data = diet, main = " ", ylab =
    "Weight Change (in kg)", xlab = "Diet", col = rgb(red =
    0.3, green = 0.3, blue = 0.3, alpha = 0.3), boxwex = 0.4)

> tapply(X = diet$WeightChange, INDEX = diet$Diet, FUN = mean)
    Atkins      LEARN      Ornish       Zone
-14.482533  -7.530623  -5.943527  -5.263537

> tapply(X = diet$WeightChange, INDEX = diet$Diet, FUN = sd)
   Atkins     LEARN    Ornish      Zone
 14.91540  13.66677  14.95681  12.62222
```

Figure 12.1 and the summary measures of the conditional weight change distributions (see Table 12.1) show that participants on all 4 diets, on average, lost weight. Figure 12.1 shows that the median values are all less than 0. The distributions are reasonably symmetric indicating the mean values are also less than 0. Participants assigned to the Atkins diet, on average, lost the greatest amount of weight. Participants on the other diets seemed to, on average, have a comparable 12-month weight change. The variation in weight change is greater than 0, and comparable across all 4 diets. In all 4 diets, the majority of participants lost weight after 12 months as indicated by the median being below 0 in each case. However, in all 4 diets, there were some participants who gained weight after 12 months.

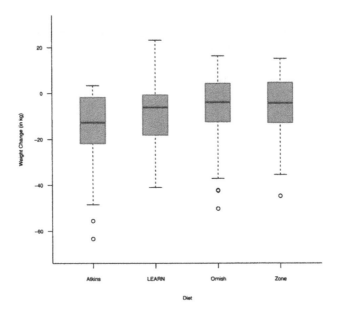

Figure 12.1: Side-by-side box-and-whiskers plots of 12-month weight change conditioned on diet.

Table 12.1: Mean and Standard Deviation for 12-month Weight Change (in kg) Conditioned on Diet.[a]

Diet	M	SD
Atkins	−14.5	14.9
LEARN	−7.5	13.7
Ornish	−5.9	15.0
Zone	−5.3	12.6

[a]The sample size for each diet was $n = 60$.

12.3 OMNIBUS TEST

The first two approaches mentioned above involve the omnibus test of group differences. The omnibus test is the simultaneous test of all possible contrasts between the groups. In other words, it is a test of any difference among the groups. Recalling the general comparison notation from the last chapter, the null hypothesis for the omnibus test is written as

$$H_0 : \Psi_k = 0 \quad \text{for all } k. \tag{12.1}$$

Another way that this null hypothesis can be written is in terms of the group means. For j groups, it is written as

$$H_0 : \mu_1 = \mu_2 = \ldots = \mu_j. \tag{12.2}$$

If all the group means are equal, the value of any contrast, simple or complex, is zero. Thus, these are equivalent ways of writing the same hypothesis.

As was pointed out in Chapter 4, differences between groups need to be evaluated relative to the variation within each group. When there are more than two groups, the differences between the groups cannot be expressed as a simple difference between the group means as when comparing two groups. With multiple groups, the difference between the groups is expressed through the variation among the group means. The premise for analyzing whether or not there are group differences is to examine the variability among the sample means (i.e., the between-group variation) relative to the variability of the individual observations within the groups (i.e., the within-group variation).

Analysis of variance (ANOVA) is a statistical method for partitioning the observed variation in a set of measurements into different components. By using this method, the total observed variation can be partitioned into a between-group component and a within-group component. The ratio of these two components is then computed and evaluated to determine whether the groups differ beyond what is expected because of either random assignment or random sampling. The justification for the analysis of variance lies in the assumption of a specific statistical model, called the *general linear model*, underlying the relationship between the observed scores and population parameters.

12.3.1 Statistical Models

Statistical models are simplified representations of relationships among variables. They can be used to model people, organizations, or any other type of social unit. In other words, all the kinds of models an educational or behavioral researcher would expect to develop or fit to data. Unlike mathematical models, statistical models are not deterministic. Consider mathematical modeling of a square geometric shape.

$$\text{Perimeter} = 4(\text{side})$$
$$\text{Area} = (\text{side})^2$$

Although one of these models is linear, and the other is nonlinear, both of these models are deterministic because *all* squares behave this way. Once the "rule" for computing the perimeter or area is known, it can always be used to fit the model to data *perfectly*. Statistical models, on the other hand, must allow for

- **Systematic components** not initially included in the model or not measured

- **Measurement error**

- **Individual variation**

The statistical model itself, represents the relationship between outcome variables and predictor variables. The goal in statistical modeling is to use a set of predictor variables to explain the total amount of variation in a set of outcome variables. Conceptually, a statistical model looks like the following:

$$\text{Outcome} = \text{Systematic Components} + \text{Residual}. \qquad (12.3)$$

The *systematic components* represent the set of predictors that are being used to explain the variation in the outcome. The *residual* represents the part of the model that allows for other systematic components not initially included in the model or not measured, and also measurement error and individual variation. Of the total variation in the outcome, the systematic part of the model more broadly represents the amount of that variation that can be explained, and the residual part more broadly represents the unexplained part of the variation.

Though this was not emphasized in the last chapter, it is true that this conceptual model underlies many of the classical statistical analyses that are used in educational and behavioral research. For example, the *t*-test, ANOVA, analysis of covariance (ANCOVA), regression analysis, factor analysis, cluster analysis, multidimensional scaling, and a plethora of other methods all can be represented with this model. The more mathematic formulation of this conceptual model, called the *general linear model*, is expressed as

$$Y_i = \underbrace{\beta_0 + \beta_1 X_1 + \beta_2 X_2 + \ldots + \beta_k X_k}_{\text{SystematicComponents}} + \underbrace{\epsilon_i}_{\text{Residual}}, \qquad (12.4)$$

where Y_i is the observed score for the ith individual. The set of β_j are the parameters that express the relationship between the set of predictor variables $(X_1, X_2, X_3, \ldots, X_k)$ and the response. This relationship is the same for all individuals; that is, the systematic portion of the model characterizes the average relationship and imposes this for each individual. Individual deviation from the average model is represented by the residual, ϵ_i.

12.3.2   Postulating a Statistical Model to Fit the Data

The variation in the observed response scores can be broken down into "explained" and "unexplained" portions. The explained portion is that part accounted for by the predictors, or the systematic components. The unexplained portion is everything left over. To break down the observed scores into explained and unexplained portions, an appropriate statistical model needs to be postulated and then fitted to the data. As such, a researcher needs to articulate the research questions in terms of outcome variables and predictors. For the example presented earlier, the research question examined by the omnibus analysis can be articulated as follows. *Does diet (predictor) explain variation in 12-month weight change (outcome variable)?*

A technical complication here is that diet is not a single predictor under the general linear model (GLM) because it is a categorical variable. Categorical variables are represented in the GLM by a series of *dummy variables*. A dummy variable is a variable whose values offer no meaningful quantitative information, but simply distinguish between categories (it is "dumb" in this respect). For example, a dummy variable called Atkins could be created to distinguish between study participants that were on the Atkins diet and those that were not.

Atkins $= 0$ if the participant was not on the Atkins diet,

Atkins $= 1$ the participant was on the Atkins diet.

When considering dummy variables, by convention,

- The variable name corresponds to the category given the value 1.

- The category given the value 0 is called the *reference* category.

One would think that in the diet example, that four dummy variables would be needed:

- A variable called Atkins to distinguish between study participants on the Atkins diet and those who are not

- A variable called LEARN to distinguish between study participants on the LEARN diet and those who were not

- A variable called Ornish to distinguish between study participants on the Ornish diet and those who were not

- And finally, a variable called Zone to distinguish between study participants on the Zone diet and those who were not

However, it can be shown that only *three* dummy variables are needed in this case. In fact, including all four of the dummy variables in the GLM will create problems for estimation with sample data. Four dummy variables are not needed because some of the information is redundant. As an example, consider the first three dummy variables just described. The three predictors Atkins, LEARN, and Ornish are mutually exclusive and exhaustive. The last dummy variable is not needed. The reason is that if an individual has a value of zero on the first three dummy variables, then it is known the individual is on the Zone diet as this is the only alternative. In general, only $j - 1$ dummy variables are required to fit a statistical model having a categorical predictor with j levels or groups. The general linear model for the omnibus analysis in this case is written as

$$\texttt{WeightChange}_i = \underbrace{\beta_0 + \beta_1(\texttt{Atkins}) + \beta_2(\texttt{Zone}) + \beta_3(\texttt{LEARN}) +}_{\text{SystematicComponents}} \underbrace{\epsilon_i}_{\text{Residual}} .$$

(12.5)

If a person is on the Atkins diet, then $\texttt{Atkins} = 1$ and all other dummy variables are zero. Therefore, the GLM for people on this diet is $\texttt{WeightChange}_i = \beta_0 + \beta_1$. If a person is on the Zone diet, then $\texttt{Zone} = 1$ and all other dummy variables are zero. The GLM for people on this diet is $\texttt{WeightChange}_i = \beta_0 + \beta_2$. Following the same logic, the GLM for people on the LEARN diet is $\texttt{WeightChange}_i = \beta_0 + \beta_3$. If a person has zero on all three dummay variables, they are on the Ornish diet. The GLM for Ornish diet people is $\texttt{WeightChange}_i = \beta_0$. In this way, different diets have different models meaning that the group means can potentially be different. The β parameters in Equation 12.5 represent group differences. The details of exactly how this works out is beyond the scope of the monograph and it is simply illustrated that this is the case in the examples below.

12.3.3 Fitting a Statistical Model to the Data

Using the observed data, sample estimates of each parameter (β) can be obtained. The general linear model is fitted in R using the $\texttt{lm()}$ function. This function takes as its argument a model formula. Model formulas in R use a concise, flexible syntax to specify statistical models, intitially proposed by Wilkinson and Rogers (1973). The tilde ($\sim$) operator is used to define a model formula. The tilde separates the outcome and the predictor variables in the model formula,

$$\text{outcome} \sim \text{predictor1} + \text{predictor2} + \dots . \tag{12.6}$$

The tilde can be read as *"is modeled by,"* so that Equation 12.6 indicates that the response variable is modeled by the predictor(s). A potential point of confusion is that dummy variables do not need to be constructed by the researcher. Rather, a single categorical grouping variable, as in the example, can be specified as the sole predictor. The grouping variable must be a factor variable (see Chapter 4). Then the $\texttt{lm()}$ function recognizes that the factor variable has multiple groups and constructs the dummy variables that are used in the GLM, but are unseen by the researcher. For example, to fit the omnibus statistical model for the diet example, the following model formula is used:

$$\texttt{diet\$WeightChange} \sim \texttt{diet\$Diet}.$$

The single $\texttt{Diet}$ variable is specified on the right-hand side, and the $\texttt{lm()}$ function will internally create the required dummy variables. The syntax to fit the omnibus model to the data and examine the output is provided in Command Snippet 12.2.

Command Snippet 12.2: Syntax to fit the omnibus model to the diet data and examine the output.

```
## Fit the omnibus model
> omnibus.model <- lm(diet$WeightChange ~ diet$Diet)

## Obtain the estimated parameters
> omnibus.model

Call:
lm(formula = diet$WeightChange ~ diet$Diet)

Coefficients:
    (Intercept)    diet$DietLEARN    diet$DietOrnish
        diet$DietZone
        -14.483            6.952            8.539
            9.219
```

These estimates can be used as a substitute for the parameters in the systematic part of the statistical model postulated earlier, namely,

$$\widehat{\text{WeightChange}}_i = -14.5 + 7.0(\text{LEARN}) + 8.5(\text{Ornish}) + 9.2(\text{Zone}). \quad (12.7)$$

Having written Equation 12.7, the meaning of the parameters and estimates in the GLM can now be discussed. Recall that above it was shown that each group has a different model. Solving the model for each group yields that group's mean value. To see this, note that the model includes three of the four levels of `Diet`. The group corresponding to the first level of the factor in alphabetical order is dropped. The hat over `WeightChange` signifies that it is a `predicted` 12-month weight change based on only the systematic portion of the model. The predicted 12-month weight change can be computed for every participant on a particular diet by substituting the appropriate values for each dummy variable.

For example, to compute the 12-month weight change for a person on the LEARN diet, the following is used:

$$\widehat{\text{WeightChange}}_i = -14.5 + 7.0(\text{LEARN}) + 8.5(\text{Ornish}) + 9.2(\text{Zone}),$$
$$\widehat{\text{WeightChange}}_i = -14.5 + 7.0(1) + 8.5(0) + 9.2(0),$$
$$\widehat{\text{WeightChange}}_i = -7.5.$$

The predicted 12-month weight change for a study participant on the LEARN diet is -7 kg, which is also the conditional mean for the group. This means that the best guess regarding change for a given person is the group average change. In a similar manner, the predicted value, or equivalently, the sample mean value can be computed for people in the other groups. The details are left to the reader who should verify that the mean for Ornish is -6 kg, and the mean for Zone is -5.3 kg.

For participants on the Atkins diet, their value on all of the dummy variables included in this model is 0. Thus, the predicted 12-month weight change for a study participant on the Atkins diet is -14.5 kg. This is directly represented in the above . equation as the estimate for β_0, or the *intercept*. The level of the categorical value that was dropped is referred to as the *reference level*. This is because the β estimate for every other level of the predictor expresses the difference between the predicted (mean) for that level and the predicted (mean) value for the reference level. For example, the β estimate associated with the LEARN diet of 7.0 indicates that a study participant on the LEARN diet has, on average, a 12-month weight change that is 7 kg more than a study participant on the Atkins diet (-7.5 kg versus -14.5 kg respectively). In this way, the mean differences between the Atkins group and each of the other groups is represented in the GLM.

12.3.4 Partitioning Variation in the Observed Scores

Recall that the goal of the omnibus model is to partition the variation in 12-month weight change between that which can be explained by differences in diet (systematic component) and that which cannot (residual). This partitioning is based on the deviation between each observed score and the marginal mean. Figure 12.2 shows this deviation for a single observed score in the data.

Figure 12.2 shows the marginal mean of -8.3 demarcated by a solid horizontal line. Four observations (weight change) are identified as solid points. The deviation between each observation and the marginal mean is shown as a solid vertical line.

Each deviation can be *decomposed* into the piece that can be explained by diet (the deviation between the marginal and conditional means) and that which cannot (the deviation between the observed value and the conditional mean). Each observed score's deviation from the marginal mean can be decomposed in a similar manner. This method is sometimes referred to as *ANOVA decomposition*.

Figure 12.3 shows the decomposition of the deviation for four observations. The marginal mean of -8.3 is again demarcated by a solid horizontal line. The conditional mean for each diet is demarcated by a dotted horizontal line. Each deviation is decomposed into two parts. The explained part of the deviation, between the marginal and conditional means, is shown as the vertical, solid line. The unexplained part of the deviation, between the conditional mean and the observation, is shown as the vertical, dashed line.

The fitted model is used to decompose the deviation for each participant's observed 12-month weight change into the explained part (i.e., predicted) and that which cannot be explained (i.e., residual). The explained part of the deviation, the portion between the marginal and conditional mean, is computed as

$$\hat{Y}_k - \bar{Y}, \tag{12.8}$$

where $\hat{Y}_k$ is the conditional mean and $\bar{Y}$ is the marginal mean. The unexplained part of the deviation, the portion between the conditional mean and the observation itself, is computed as

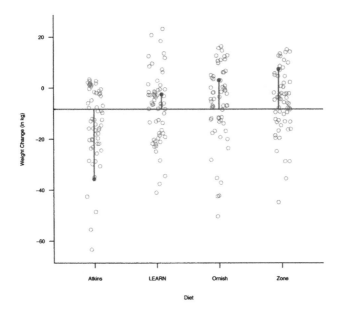

Figure 12.2: The deviation between the observed score and the marginal mean (horizontal line) for four observations.

$$Y_i - \hat{Y}_k, \tag{12.9}$$

where Y_i is the ith observation and $\hat{Y}_k$ is the conditional mean. This is called the *residual*. Equation 12.10 shows the decomposition for the 163rd diet study participant.

$$
\begin{aligned}
Y_{163} - \bar{Y} &= (\hat{Y}_{163} - \bar{Y}) + (Y_{163} - \hat{Y}_{163}), \\
3.1 + 8.3 &= (-5.9 + 8.3) + (3.1 + 5.9), \\
11.4 &= 2.4 + 9.0.
\end{aligned}
\tag{12.10}
$$

This decomposition is carried out on all 240 deviations. The decomposed deviations can be used to study the amount of explained variation in the data. The total variation in a set of scores is expressed as the sum of squared deviations from a mean value. This is referred to as a *sum of squares*. The equation for the sum of squares for the response is

$$SS_Y = \sum \left(Y_i - \bar{Y}\right)^2 \tag{12.11}$$

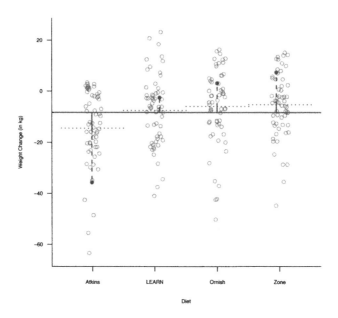

Figure 12.3: The deviation between the observed score and the marginal mean (horizontal line) for four observations. Each has now been decomposed into the part of the deviation that can be explained by diet (solid) and that which cannot (dashed).

where Y_i is the observed value for the ith individual, and $\bar{Y}$ is the mean value for all the Y_i, that is, the marginal mean. For example, the total variation in the observed 12-month weight changes can be found using

$$\text{SS}_{\text{WeightChange}} = \sum \left(\text{WeightChange}_i + \overline{\text{WeightChange}} \right)^2$$

A visual representation of the squared deviations from the marginal mean weight change for an observation from each diet is shown in Figure 12.4. The area of each square represents the squared deviation for a single observation. If this was carried out for each of the 240 observations and these squared deviations (i.e., areas) were summed, that would be the sum of squares for the total variation in the data for 12-month weight change.

As pointed out in Chapter 11, the sum of squares for the response is the same quantity in the numerator of the sample variance. Using this information, SS_Y can be computed using the syntax in Command Snippet 12.3.

The lm() object omnibus.model has computed and stored the predicted values and the residuals for each of the 240 observations based on the decomposition specified in the model. The 240 predicted values are stored in a component

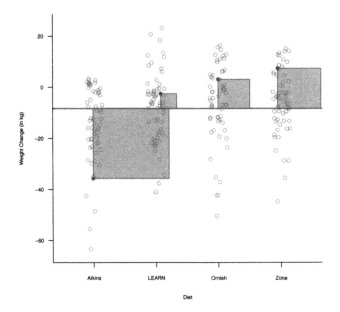

Figure 12.4: Graphical representation of the squared deviations from the marginal mean weight change for four observations.

called `fitted.values`, and the 240 residuals are stored in a component called `residuals`. These can be accessed using either `omnibus.model$fitted.values` or `omnibus.model$residuals`, respectively.

Command Snippet 12.3: Syntax to compute the total variation in `WeightChange` via the sum of squares.

```
> var(diet$WeightChange) * (240 - 1)
[1] 49959.62
```

After decomposing each participant's observed score, the amount of variation that is not explained by diet is partitioned into the residual value. As pointed out, the variation in a set of scores is expressed as the sum of squared deviations from a mean value. The residuals are a deviation from the conditional mean (predicted) value. A graphical representation of the variation based on the decomposed deviations is shown in Figures 12.5 and 12.6.

Figure 12.5 shows the squared decomposed model (explained) deviations for four observations. The area of each square represents the squared deviation between the conditional and marginal mean. Figure 12.6 shows the squared decomposed residual

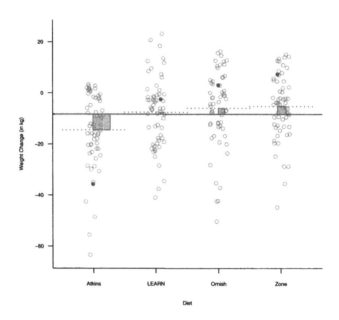

Figure 12.5: Graphical representation of the squared decomposed model deviations for four observations—the deviation between the conditional mean (dotted horizontal line) and the marginal mean (solid horizontal line). The squares represent the squared model deviations for the four observations.

(unexplained) deviations for four observations. The area of each square represents the squared deviation between the conditional mean and the observation.

To find the variation that is unexplained by diet, the squared residuals are summed using

$$ \text{SS}_{\text{Residual}} = \sum \left(\hat{\epsilon}_i \right)^2 $$

This can be computed using the syntax in Command Snippet 12.6. Lastly, the variation explained by diet is the leftover variation after removing the residual variation from the total observed variation. This is also computed in Command Snippet 12.4.

The two measures of variation, the explained and unexplained, are often turned into proportions based on the total variation. The proportion of the total variation that can be explained by the predictors is the estimate of the omnibus measure of effect $\hat{\eta}^2$.

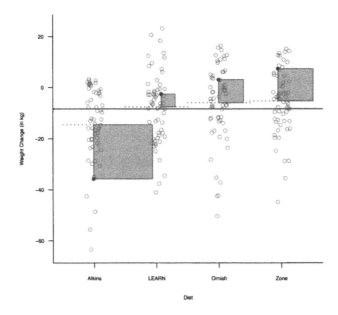

Figure 12.6: Graphical representation of the squared decomposed residual deviations for four observations—the deviation between the conditional mean (dotted horizontal line) and the observed value. The squares represent the squared residual deviations for the four observations.

Command Snippet 12.4: Syntax to compute the variation in `residuals` via the sum of squares.

```
## Compute the residual variation
> sum(omnibus.model$residuals ^ 2)
[1] 46744.3

## Compute the variation explained by diet
> 49959.62 - 46744.3
[1] 3215.32
```

$$\hat{\eta}^2 = \frac{3215}{49,960}$$
$$= 0.064$$

Thus, of the total variation in the observed 12-month weight change, differences in diet explain roughly 6% of that variation and roughly 94% remains unexplained due to other systematic components not initially included in the model or not measured, and measurement error or individual variation.

12.3.5   Randomization Test for the Omnibus Hypothesis

In considering the omnibus effect, the inferential question is whether or not the effect is bigger than would be expected because of the random assignment to diets. In order to determine this, the null hypothesis of no effect is assumed, that is,

H_0 : There is no effect of diet on 12-month weight change.

This is equivalent to saying the weight loss means of all the groups are equal. Under this hypothesis, the observed 12-month weight change measurements could be permuted since the diet makes no difference on the participants' weight change. To carry out a randomization test to obtain the distribution for the omnibus effect under the null hypothesis, the steps presented in Figure 12.7 are used.

Randomization Test for the Omnibus

- Randomly permute the observed sample data.

- Compute $\hat{\eta}^2$—the estimated effect from the permuted data.

- Repeat these first two steps many times, say R times, each time recording the estimated contrast $\hat{\eta}^2$.

- The distribution of $\hat{\eta}_1^2, \hat{\eta}_2^2, \hat{\eta}_3^2, \ldots, \hat{\eta}_R^2$ can be used as an estimate of the sampling distribution of η^2 under the null model. The value of $\hat{\eta}^2$ from the observed data can be evaluated using this distribution.

Figure 12.7: Steps to carry out a randomization test to obtain the distribution for the omnibus effect under the null hypothesis.

Command Snippet 12.5 shows the syntax for carrying out the randomization test based on the null hypothesis of no effect. When testing hypotheses for models, the permutations can be drawn in the lm() function. The sample() function is used within the lm() function to permute the response scores. The residuals based on the permuted data are then accessed. This is all completed in a single chained computation which makes up the expr= argument of the replicate() function. Based on these results, there is very strong evidence against the null hypothesis of no effect ($p = 0.002$).

Command Snippet 12.5: Set of commands to carry out the randomization test for the first contrast.

```
## Permute the data within the model
> permuted <- replicate(n = 4999, expr =
    lm(sample(diet$WeightChange) ~ diet$Diet)$residuals)

## Function to compute the estimated effect size
> eta.squared.omnibus <- function(data) {
  ss.total <- 49959.62
  ss.residual <- sum(data ^ 2)
  ss.diet <- ss.total - ss.residual
  ss.diet / ss.total
  }

## Apply the effect size to each of the 4999 permuted samples
> perms <- apply(X = permuted, MARGIN = 2, FUN =
    eta.squared.omnibus)

## Calculate the Monte Carlo p-value
> length(perms[abs(perms) >= 0.064])
[1] 8
> (8 + 1) / (4999 + 1)
[1] 0.0018
```

12.4 GROUP COMPARISONS AFTER THE OMNIBUS TEST

If researchers find strong evidence against the null hypothesis for the omnibus test, then many go on to test further contrasts to explore how the groups are different. While any contrast can be tested, a common approach is to test all the simple or pairwise contrasts. Recall from earlier in this chapter that pairwise contrasts examine the mean difference between exactly two groups at a time.

In the analysis of the diet data, there are six pairwise contrasts that could be tested: that is, Atkins versus LEARN, Atkins versus Ornish, etc. In general, with j groups, there are $j(j-1)/2$ pairwise comparisons. Each of the null hypotheses associated with the six pairwise contrasts is presented below.

$$H_{01} : \mu_{Atkins} - \mu_{Zone} = 0$$
$$H_{02} : \mu_{Atkins} - \mu_{LEARN} = 0$$
$$H_{03} : \mu_{Atkins} - \mu_{Ornish} = 0$$
$$H_{04} : \mu_{Zone} - \mu_{LEARN} = 0$$
$$H_{05} : \mu_{Zone} - \mu_{Ornish} = 0$$
$$H_{06} : \mu_{LEARN} - \mu_{Ornish} = 0$$

These contrasts are tested in the exact same manner as they are in planned comparisons. Table 12.2 presents the contrast coefficients, observed contrast value, randomization test results, eta-squared estimate, and bootstrap results for the interval estimate of eta-squared for each of the six pairwise contrasts. The syntax is not presented, but follows along the same lines as the syntax presented earlier for the first planned contrast.

After the pairwise contrast tests are carried out, there are two main schools of thought about whether these p-values should be reported as they are, or whether the p-values can be reported as is in Table 12.2, or whether they should be adjusted for the fact that multiple unplanned tests are performed. Methods of adjustment invariably adjust the p-values *upwards*, making them larger than unadjusted p-values. When an unadjusted p-value is very small or very large, rarely will an adjustment change the judgment regarding statistical reliability. It is only for those unadjusted p-values that hover around thresholds of statistical reliability, such as 0.05, that adjustment might lead to a different judgment than under unadjustment.

12.5 ENSEMBLE-ADJUSTED p-VALUES

In exploratory analyses, after the omnibus test is performed, there are some re-searchers who may want to adjust their p-values based on the number of contrast tests that were carried out. Rosenthal and Rubin (1983, p. 540) refer to this as computing an "ensemble-adjusted p-value." The rationale behind ensemble adjustment is that the contrast in question are unplanned. This implies the researcher does not know if any of the null hypotheses evaluated will be rejected. It can be shown that if all the null hypotheses being evaluated are true, then the false-positive (type I error) rate of the collection of statistical tests will be higher than any one test. This is analogous to walking into a mine field with only one mine as opposed to many. Suppose a mine represents a false-positive result and a hypothesis tests is analogous to root around in the mine field. If there is only one mine, there is a much smaller chance of setting it off when exploring the mine field than when there are multiple mines. Ensemble-adjusted p-values act as special protection in the case of multiple mines in the field. The method attempt to limit the possible triggering of a mine to the case when only one mine is in the field.

One of the most common methods for computing ensemble-adjusted p-values, and also one of the straightforward methods, is the Dunn–Bonferroni (Dunn, 1961) adjustment. This method is presented here for pedagogical purposes, but the pro-cedure is not generally recommended due to the conservative p-values it produces, especially with large numbers of contrasts. The Dunn–Bonferroni method adjusts the p-value based on the number of contrasts being tested using,

$$p_{\text{adj.}} = p_k \times m, \qquad (12.12)$$

where p_k is the unadjusted p-value obtained from the randomization or bootstrap test for the kth contrast, and m is the number of contrasts being tested. Using the Monte Carlo p-values from each of the nonparametric bootstrap tests, the Dunn–Bonferroni adjusted p-values are computed in Command Snippet 12.6. The p-values in the output

Table 12.2: Contrast Coefficients, Observed Contrast Value, Randomization Test Results, Eta-Squared Estimate, and Bootstrap Results for Interval Estimate of Eta-Squared for Each of Six Pairwise Contrasts[a]

Comparison	Contrast Coefficients	$\hat{\Psi}$	*p*-Value	$\hat{\eta}^2$	Bootstrap Interval
Atkins vs. Zone	[1, 0, 0, −1]	−9.2	0.0008	0.510	[0.012, 0.126]
Atkins vs. Ornish	[1, 0, −1, 0]	−8.5	0.0010	0.044	[0.007, 0.121]
Atkins vs. LEARN	[1, −1, 0, 0]	−7.0	0.0092	0.029	[0.003, 0.090]
Zone vs. LEARN	[0, −1, 0, 1]	2.3	0.3896	0.003	[0.000, 0.026]
LEARN vs. Ornish	[0, 1, −1, 0]	−1.6	0.5374	0.002	[0.000, 0.016]
Zone vs. Ornish	[0, 0, −1, 1]	0.7	0.7896	0.000	[0.000, 0.003]

[a]The *p*-value is based on the randomization test using 4999 permutations of the data. The bootstrap interval is a bias-corrected-and-accelerated adjusted interval based on a nonparametric bootstrap using 4999 replicate data sets.

are interpreted with the same standard as mentioned previously. Most commonly, the judgments of effect size in Table 8.2 are used.

Command Snippet 12.6: Computation of the ensemble-adjusted p-values for each of the six pairwise contrasts.

```
## Create a vector of the unadjusted p-values
> p.unadjusted = c(0.0008, 0.0010, 0.0092, 0.3896, 0.5374,
    0.7896)

## Perform the Dunn-Bonferroni adjustment
> p.unadjusted * 6
[1] 0.0048 0.0060 0.0552 2.3376 3.2244 4.7376
```

Consistent with what was previously mentioned, each ensemble-adjusted p-value is larger than the associated unadjusted p-value. The problem with the Dunn–Bonferroni method is that it tends to overadjust, making the prospect of rejecting an null hypothesis more difficult than it should be. Another comment regarding the output in Command Snippet 12.6 is that some of the ensemble-adjusted p-values are greater than 1. In theory this is not possible and computer programs typically set 1 as the upper boundary.

There are many ensemble adjustment methods that are superior to the Dunn–Bonferroni method, in the sense that they do not overadjust. These methods have various underlying approaches. For example, some use simultaneous adjustment, where all of the p-values are adjusted at the same time, while others use sequential adjustment, where the adjustment method begins with either the most or least extreme p-value, and then a series of successive adjustments is made on the remaining p-values in turn [e.g., the Holm method (Holm, 1979)]. Some methods are more conservative (i.e., producing higher p-values) than others [e.g., the Scheffé method (Scheffé, 1953)]. Hochberg and Tamhane (1987) and Hsu (1996) offer detailed descriptions of many ensemble adjustments.

12.5.1 False Discovery Rate

False discovery rate (FDR) is a relatively new approach to the multiple comparisons problem. Instead of controlling the chance of at least one type I error, FDR controls the expected proportion of type I errors. Methods of testing based on the FDR are more powerful than methods based on the familywise error rate simply because they use a more lenient metric for type I errors. Less technically speaking, the FDR methods are less prone to overadjustment than many other methods, especially the Dunn–Bonferonni method. However, if there is truly no difference in the groups, a FDR-controlling method has the same control as the more conventional methods.

One method for controlling the FDR is called the Benjamini–Hochberg procedure (Benjamini & Hochberg, 1995). Since its introduction, there has been a growing pool of evidence showing that this method may be the best solution to the multiple comparisons problem in many practical situations (Williams, Jones, & Tukey, 1999).

Because of its usefulness, the Institute of Education Sciences has recommended this procedure for use in its *What Works Clearinghouse* handbook of standards (Institute of Education Sciences, 2008).

The Benjamani–Hochberg adjustment to the *p*-value can be implemented using the p.adjust() function.[1] This function takes two arguments to make ensemble adjustments. The first, p=, provides the function a vector of unadjusted *p*-values. The second argument, method=, takes a character string that identifies the adjustment method to use. The Benjamani–Hochberg adjustment is carried out using the argument method="BH". Command Snippet 12.7 shows the use of this function to obtain the Benjamani–Hochberg ensemble-adjusted *p*-values for the pairwise contrasts tested in the diet analysis. Of interest is that the ensemble-adjusted *p*-values in the output are larger than the unadjusted versions (except for the largest one, which does not change). However, the adjustment is not as severe as with the Dunn–Bonferroni. The FDR method strikes a good balance between protection from a false positive and the ability to reject a false null hypothesis.

Command Snippet 12.7: Computation of the Benjamani–Hochberg ensemble-adjusted *p*-values for each of the six pairwise contrasts.

```
> p.adjust(p = p.unadjusted, method = "BH")
[1] 0.00300 0.00300   0.01840 0.58440 0.64488 0.78960
```

12.6 STRENGTHS AND LIMITATIONS OF THE FOUR APPROACHES

For any researcher interested in drawing inferences about group differences, there are key analytic decision points that must be made at different stages of an analysis. One initial consideration is the analytic framework used to compare the groups. Chapters 11 and 12 were both framed in terms of four common approaches typically used by educational and behavioral science practitioners when it comes to comparing multiple groups. They are: (1) planned comparisons; (2) unplanned, but adjusted group comparisons without the omnibus test; (3) the omnibus test followed by unadjusted group comparisons; and (4) the omnibus test followed by adjusted group comparisons. Strengths and weaknesses of all four approaches are examined in the context of the ongoing controversy surrounding these analytic paths.

12.6.1 Planned Comparisons

Planned comparisons are confirmatory in nature and as such often reflect the researcher's substantive knowledge and theoretical work in the field. If orthogonal contrasts are planned in advance, contemporary practice favors adopting unadjusted

---

[1]This function also implements the Holm (1979), Hochberg (1988), Hommel (1988), Dunn (1961), and Benjamini and Yekutieli (2001) adjustments.

p-values; meaning testing each hypothesis at the nominal (i.e., theoretical) level of statistical reliability, usually 0.05. For planned, but nonorthogonal contrasts, opinions differ as to whether an adjustment should be made. One line of thought is that nonorthogonal contrasts involve redundant information; and the outcome of one test is not independent of those for other tests. In such cases, some method of adjustment is typically advocated. In fact, the degree of redundancy among the contrasts is the basis for many different *p*-value adjustment procedures (Hancock & Klockars, 1996). The other perspective is to use unadjusted *p*-values for the small number of theoretically driven hypothesis tests that are performed (Keppel & Wickens, 2007). This approach is more likely to detect true population differences.

12.6.2 Omnibus Test Followed by Unadjusted Group Comparisons

When no hypotheses are stated in advance of examining the data, then the analysis is exploratory in nature. As an initial exploratory step, researchers have historically examined the simultaneous differences among group means. If the primary research question focuses on detecting omnibus mean differences among groups, then the analysis of variance is the statistical vehicle to answer this question. If on the other hand, the omnibus test is used as a ritual to performing any subsequent follow-up group mean comparisons—for example, as a method of controlling the type I error rate—contemporary wisdom suggests that this course of action may be unnecessary. This perspective runs counter to R. A. Fisher's (1935) view that follow-up multiple comparisons were permissible only after rejecting the omnibus hypothesis. Rejection of the omnibus hypothesis provided the requisite type I error protection such that no adjustments were needed for subsequent hypothesis tests. The broader question about whether or not to adjust the *p*-values comes from the question of type I error control and the value the researcher places on this.

A number of authors have advocated that no adjustment for the multiple group comparisons is necessary (e.g., Rothman, 1990; Saville, 1990). Their arguments are varied, but essentially boil down to supporting a per-comparison approach to controlling type I errors. This will lead to more type I errors when the complete null hypothesis is true (i.e., $\mu_1 = \mu_2 = \cdots = \mu_g$). Yet, they argue that this scenario is almost never true and therefore, in those cases in which the hypothesis is false, the chance of committing a type II error (incorrectly retaining a false null hypothesis) necessarily increases. For some researchers, committing a type II error is seen as more egregious than that of committing a type I error. Thus, unadjusted comparisons are attractive to them. In support of this position, Hochberg and Tamhane (1987, p. 6) note that when "inferences are unrelated in terms of their content or intended use (although they may be statistically dependent), then they should be treated separately and not jointly."

12.6.3 Omnibus Test Followed by Adjusted Group Comparisons

In the context of the omnibus test, the alternative hypothesis allows for inequality among the group means, and if the null hypothesis is eventually rejected, all that can be said is that there is some type of linear contrast of the group means that is non-zero. A conventional analytic strategy upon rejecting the omnibus null hypothesis would be to conduct further hypothesis tests corresponding to contrasts that are of particular interest or those suggested by the data. Researchers that worry about type I error control in these unplanned multiple tests favor some sort of p-value adjustment. The looming issue is that the type I error rate will become inflated due to the number of tests within a family of tests—assuming that the null hypotheses within this family are true. For example, if there are six comparisons to make, under the guise that all six null hypotheses are true, then the probability of making at least one type I error when testing all six contrasts is not 0.05 but a whopping 0.26! Rothman (1990) points out that adjustment for multiple comparisons is like an insurance policy against mistakenly rejecting a null hypothesis when the null is actually true. These adjustments usually involve increasing the p-value, thus making it more difficult to reject any one hypothesis.

Regrettably, the cost of the insurance policy is to increase the frequency of an incorrect statement (or conclusion) that asserts there is no mean difference when in reality a true difference exists. Because of this, there is some dissatisfaction for these adjustments in general, and in particular for multiple testing stemming from an exploratory analysis. As Tukey (1953) suggests, exploratory analyses are for generating hypotheses for future research. If the analysis is exploratory, why not run all the comparisons and rank-order them in terms of their unadjusted p-values? This will certainly help indicate which hypotheses may be useful to examine in future research. Modern multiple comparison procedures, like those mentioned in this chapter, acknowledge the trade-off between statistical power and control over type I errors. They attempt to hold the familywise error rate at the nominal level while providing the greatest chance of finding a true difference if it exists.

12.6.4 Adjusted Group Comparisons without the Omnibus Test

Some controversy exists regarding the question of whether one should even run the omnibus test on groups before embarking on comparisons that seem most interesting. A number of authors (Cohen, 1994; Hancock & Klockars, 1996; Rosnow & Rosenthal, 1996; Serlin & Lapsley, 1993) have criticized the efficacy of the omnibus test of this null hypothesis. In their critiques several compelling reasons have been provided of why researchers should avoid the omnibus testing all together. First and foremost, it is not a very compelling hypothesis. Knowing that some type of mean difference exists among groups is not particularly interesting as it seems obvious to be the case in many research contexts. It could be argued that it is better to bypass the omnibus test altogether and jump right to the comparisons suggested by the data. As Games (1971, p. 560) stated, "There seems to be little point in applying the overall [omnibus] test prior to running c contrasts by procedures that set [the familywise error rate] $\leq \alpha \ldots$ if the c contrasts express the experimental interest directly, they are justified whether the overall [omnibus test] is significant or not and [familywise error rate] is still controlled." Secondly, the term unplanned or post-hoc frequently is misinterpreted as the follow-up to an omnibus test. Researchers are frequently

Table 12.3: Randomization Test Results for Both Unadjusted Method and Benjamani–Hochberg Ensemble-Adjusted Method[a]

Comparison	Unadjusted	Ensemble Adjusted
Atkins vs. Zone	0.0008	0.0030
Atkins vs. Ornish	0.0010	0.0030
Atkins vs. LEARN	0.0092	0.0184
Zone vs. LEARN	0.3896	0.5844
LEARN vs. Ornish	0.5374	0.6449
Zone vs. Ornish	0.7896	0.7896

[a] The p-value for both methods is based on the randomization test using 4999 permutations of the data.

encouraged to conduct multiple comparisons only after rejecting the null hypothesis of the omnibus test, the rationale being that it provides protection from the perils of inflated type I errors (see, e.g., Hinkle, Wiersma, & Jurs, 2003). The supposition that an omnibus test offers protection is not, however, completely accurate. In the situation when the omnibus null hypothesis is false, the conventional analysis of variance does not maintain control over the type I error rate for the family of subsequent comparisons (Hancock & Klockars, 1996). In addition, many procedures that are used to adjust p-values, like those included in this chapter, control the type I error rate all by themselves making the omnibus test redundant. In fact, this redundancy could potentially hinder the probability of detecting real differences by unnecessarily increasing the probability of a type II error.

12.6.5 Final Thoughts

There are many potential decision points in an analysis of comparing groups. This monograph has presented four popular methods of testing group mean differences, each having their own advantages as well as limitations. In the end, it is the responsibility of every researcher to provide a rationale for the methodological choices that he or she makes. This rationale should be based on sound, principled statistical and/or theoretical arguments in favor of one path over another.

12.7 SUMMARIZING THE RESULTS OF UNPLANNED CONTRAST TESTS FOR PUBLICATION

An example write-up for the unplanned contrast results is provided below. As with any contrast analysis, it is important to indicate the approach that was taken. For example, a researcher using the omnibus test and then adjusting the p-value for the pairwise contrasts might provide the following write-up.

Sample Write-Up

Twelve-month weight change in overweight, nondiabetic, premenopausal women was analyzed to examine whether there were differences between four diets. A randomization test was used to examine the omnibus null hypothesis of no diet differences. A Monte Carlo corrected p-value (Davison & Hinkley, 1997) for this test was computed by permuting the data 4999 times. The results of the test indicated that these data provide moderate evidence (see Efron & Gous, 2001) against the null hypothesis of no difference ($\hat{\eta}^2 = 0.064, p = 0.039$).

These differences in diet were further analyzed by testing all six pairwise contrasts. Each contrast was tested using a randomization test based on 4999 permutations of the data. The Monte Carlo p-values from these contrasts were adjusted using an ensemble adjustment method proposed by Benjamini and Hochberg (1995). Based on these results, there is very strong to overwhelming evidence of a difference between the Atkins and Zone diets ($p = 0.003$), as well as between the Atkins and Ornish diets ($p = 0.003$). There is also substantial evidence of a difference between the Atkins and LEARN diets ($p = 0.018$). There was no evidence to suggest any other differences based on the results of the other pairwise contrasts. Table 12.3 presents both the unadjusted and ensemble-adjusted results.

12.8   EXTENSION: PLOTS OF THE UNPLANNED CONTRASTS

Researchers often produce plots to convey information about the group differences. One plot that is commonly reported is the *dynamite plot*. An example of a dynamite plot that a researcher might produce for the weight change data is shown in Figure 12.8. Dynamite plots, so called because of their resemblance to the dynamite detonator used in cartoons, represent the conditional means via the height of the bar and the standard error of the mean in the length of the whisker. Figure 12.8 shows a distinct difference between the Atkins diet and the three other diets. The difference between Atkins and LEARN appears to not be attributable to sampling error. The standard error "detonator" atop the LEARN bar does not overlap with the top of the Atkins bar. Apart from the Atkins diet, there is overlap among the standard error bars of all the other diets, suggesting they may have equal means within sampling error.

The dynamite plot is relatively primitive in that it can be difficult to determine precisely which group means are different. For example, in Figure 12.8, the error bar for the Atkins diet does not descend below the top of its solid bar. Therefore, it cannot be determined precisely from the plot if the Atkins group mean has a statistically reliable difference compared to the LEARN mean. In fact, Koyama (n.d.) suggests that dynamite plots are not helpful, pointing out that such plots hide the actual data in each group, masking important features of each distribution such as sample size and variation. More importantly, this display does not easily allow consumers of the

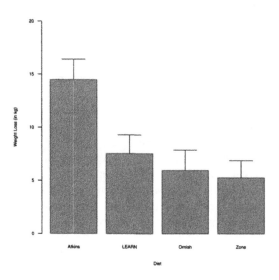

Figure 12.8: Dynamite plot showing the conditional mean weight change and standard error for each diet.

research to identify group differences, except in extreme cases, such as the Atkins diet in Figure 12.8. A better graphical presentation of the differences examined using the unplanned contrasts can be created by plotting the bootstrap interval of the mean difference for each of the six contrasts.

To create this plot, three vectors are created. These vectors contain the point estimates, the lower bootstrap limits, and the upper bootstrap limits for the contrasts. Command Snippet 12.8 shows the creation of these three vectors based on the diet analysis.

Command Snippet 12.8: Create vectors of the point estimates, lower bootstrap limits, and upper bootstrap limits.

```
## Vector of point estimates
> point.estimate <- c(-9.2, -8.5, -7.0, -2.3, -1.6, 0.7)

## Vector of lower limits
> lower.limit <- c(-14.2, -13.9, -12.2, -6.9, -6.3, -5.8)

## Vector of upper limits
> upper.limit <- c(-4.3, -3.1, -2.1, 2.6, 3.8, 4.1)
```

The `plot()` function is used to set up the plot as shown in Command Snippet 12.9. Recall that the argument `type="n"` draws the plot according to the graphing parameters set in the function, but does not plot any results. Inclusion of the argument yaxt="n" will omit the y-axis. Even though the axis itself is not included, the argument ylim= is still included to set the limits of the plotting area to correspond to the number of contrasts.

Command Snippet 12.9: Set up the plotting area for the plot of the bootstrap intervals of the mean difference for the six contrasts.

```
> plot(x = point.estimate, xlab = "Contrast Value", ylab = " ",
    xlim = c(-30, 10), ylim = c(1, 6), type = "n", bty = "n",
    yaxt = "n")
```

The bootstrap interval for each contrast can then be added to the plot using the segments() function. This function adds a line segment to an existing plot. It takes the arguments x0=, y0=, x1=, and y1= to indicate the starting coordinates (x_0, y_0) and ending coordinates (x_1, y_1) for the line segment to be drawn. Additional arguments, such as col=, lty=, etc., can also be included in this function.

Command Snippet 12.10: Syntax to draw the six bootstrap intervals.

```
> segments(x0 = lower.bound, y0 = 1:6, x1 = upper.bound, y1 =
    1:6, lty = "dotted")
```

Rather than type this function six times, one for each contrast, the vectors created earlier are used in the function as shown in Command Snippet 12.10. Note that the first four arguments, indicating the coordinates, are each composed of a vector having six elements, respectively. The lty= argument only has one element. This is because R uses a *recycling* rule that extends any vector that is shorter by "recycling" the elements until it has the same length. Thus, because of the recycling, it is equivalent to using lty = c("dotted", "dotted", "dotted", "dotted", "dotted", "dotted").

Lastly, the point estimate of the mean difference for each contrast is added to the plot as a solid point using the points() function. This function takes the arguments x= and y= to indicate the coordinates at which the point is to be drawn. The argument pch= is also included to change the default plotting character from an open circle to a solid point. This is shown in Command Snippet 12.11.

Labels are also added to the plot to identify the contrasts using the text() function, also shown in Command Snippet 12.11. Figure 12.9 shows the resulting plot of the estimates of the mean difference for each contrast based on 4999 bootstrap replicates. To aid interpretation, a vertical line is plotted at zero, the parameter value associated with no difference. The contrasts that show strong evidence of a difference in 12-month weight change between the two diets based on the interval estimates have also been given a different line type. This plot is congruent with the results of the analysis. In contrast to the dynamite plot, Figure 12.9 depicts the bootstrap intervals about the means. Thus, the intervals can be inspected to see if 0 falls within the bounds. If so, the groups do not have a statistically reliable difference. If 0 is not in the bootstrap interval, the difference is statistically reliable. Inspection of Figure 12.9 shows that the bottom three group comparison intervals do not contain 0. Therefore, the bottom three comparisons are statistically reliable, whereas the top three are not.

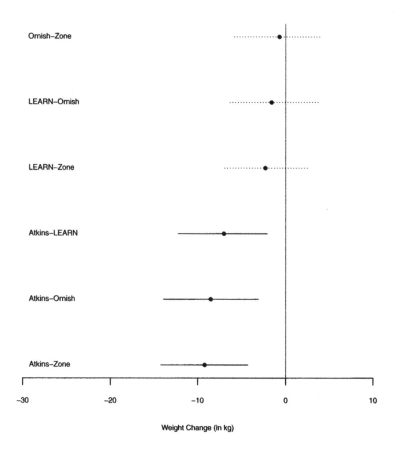

Figure 12.9: Plot showing the point estimate and bootstrap interval estimate of the mean difference for each contrast based on the bias-corrected-and-accelerated adjustment using 4999 bootstrap replicates.

12.8.1 Simultaneous Intervals

Like the p-values produced for these analyses, the bootstrap interval estimates can be unadjusted or adjusted. If the researcher has used an ensemble adjustment method, then it is also appropriate to provide an adjusted interval estimate. Adjusted interval estimates are referred to as *simultaneous intervals*. The adjustment for the interval estimate is carried out by including the argument `conf=` in the `boot.ci()` function. The value that this argument takes is

$$1 - \alpha^{\star}, \tag{12.13}$$

where, $\alpha^{\star}$ is the ensemble-adjusted significance level. The Benjamani–Hochberg procedure uses a different adjusted significance level for each contrast. In order to find the ensemble-adjusted significance levels, the contrasts are ordered, based on their *unadjusted p*-values, from most evidence against the null hypothesis (smallest) to least evidence against the null hypothesis (largest). Then, the ensemble-adjusted significance level is computed by dividing the position in the sequence by the number of contrasts and multiplying this quantity by an overall significance level. Table 12.4 shows each contrast and the adjusted level of significance.

Command Snippet 12.11: Syntax to add the point estimate for each contrast to the plot.

```
## Add the point estimates to the plot
> points(x = point.estimate , y = 1:6, pch = 20)

## Label each contrast
> text(x = -30, y = 1:6, label = c("Atkins-Zone",
    "Atkins-Ornish", "Atkins-LEARN", "LEARN-Zone",
    "LEARN-Ornish", "Ornish-Zone"), cex = 0.8, pos = 4)

## Add vertical line at 0
> abline(v = 0)
```

The quantity $1 - \alpha^{\star}$ can then be used in the optional `conf=` argument for the `boot.ci()` function. For example, to compute the BC_a interval limits for the contrast to compare the Atkins and Ornish diets, the argument `conf=0.983` is added to the `boot.ci()` function. It is noted that for readers interested in a computing challenge, a function could be written to more easily compute the ensemble-adjusted significance levels. The unadjusted and adjusted (simultaneous) interval estimates for the unplanned pairwise contrasts are presented in Table 12.5.

Figure 12.10 shows a plot of the simultaneous intervals for each of the six contrasts based on the Benjamani–Hochberg adjusted significance level. To aid interpretation, a vertical line is plotted at zero, the parameter value associated with no difference. The contrasts that show strong evidence of a difference in 12-month weight change between the two diets based on the interval estimates have a different line type.

Table 12.4: Computation of Adjusted Significance Level for Benjamani–Hochberg Simultaneous Confidence Intervals Based on Sequential Unadjusted p-Values

Adjusted Comparison	p-Value	Position	$\alpha^\star$
Atkins vs. Zone	0.0008	1	$1/6 \times 0.05 = 0.008$
Atkins vs. Ornish	0.0010	2	$2/6 \times 0.05 = 0.017$
Atkins vs. LEARN	0.0092	3	$3/6 \times 0.05 = 0.025$
Zone vs. LEARN	0.3896	4	$4/6 \times 0.05 = 0.033$
LEARN vs. Ornish	0.5374	5	$5/6 \times 0.05 = 0.042$
Zone vs. Ornish	0.7896	6	$6/6 \times 0.05 = 0.050$

Table 12.5: Unadjusted and Simultaneous Bootstrap Intervals for Each of Six Contrasts[a]

Comparison	Unadjusted Interval	Simultaneous Interval
Atkins vs. Zone	$[-14.2, -4.3]$	$[-16.1, -2.7]$
Atkins vs. Ornish	$[-13.9, -3.1]$	$[-15.1, -1.8]$
Atkins vs. LEARN	$[-12.2, -2.1]$	$[-13.2, -1.2]$
Zone vs. LEARN	$[-6.9, 2.6]$	$[-7.3, 3.0]$
LEARN vs. Ornish	$[-6.3, 3.8]$	$[-6.5, 4.0]$
Zone vs. Ornish	$[-5.8, 4.1]$	$[-5.8, 4.1]$

[a]The simultaneous interval is computed based on the Benjamani–Hochberg adjusted significance level. Each bootstrap interval is a bias-corrected-and-accelerated adjusted interval based on a nonparametric bootstrap using 4999 replicate data sets.

12.9   FURTHER READING

The examination and testing of multiple comparisons has a long history, being discussed as early as 1843 when the problem was identified as a result of the exploration of subpopulation differences all within a single population (Cournot, 1843/1984). Since that time numerous methods for simultaneous inference have been introduced. Of the more promising approaches beyond those discussed in this chapter is a pairwise comparison procedure based on information criteria computed within a maximum likelihood estimation framework (Dayton, 1998, 2003) and best subsets approach within this same framework (Dayton & Pan, 2005). Westfall (1997) demonstrated how to bootstrap adjusted p-values directly. This interesting approach allows mod-

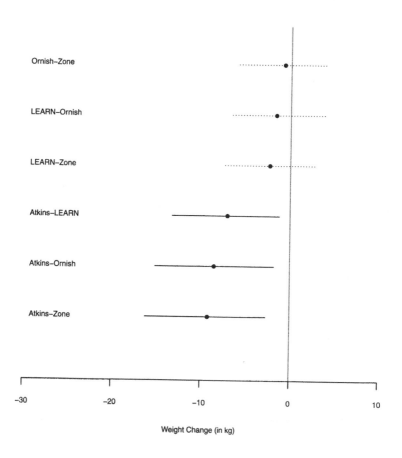

Figure 12.10: Plot showing the point estimate and bootstrap interval estimate of the mean difference for each contrast based on the Benjamani–Hochberg adjusted significance level. Each interval uses the bias-corrected-and-accelerated adjustment using 4999 bootstrap replicates. A vertical line is plotted at zero, the parameter value associated with no difference.

els corresponding to mean comparisons to be specified to correspond to empirically based distributional assumptions as well as variance equality. Gelman, Carlin, Stern, and Rubin (2003) discuss the multiple comparison problem within a Bayesian and multilevel modeling framework. Several books and papers over the years have comprehensively examined many of these procedures. Hochberg and Tamhane (1987), Westfall and Young (1993), and Shaffer (1995) provide good starting points. Bretz, Hothorn, and Westfall (2011) present many of these methods in R under a more classical framework. Lastly, it is noted that the debate over whether to use ensemble p-values and simultaneous confidence limits when making multiple comparisons—especially when those comparisons are hypothesized a priori—is still an active debate. Many methodologists have provided good reasons for not adjusting inferences when multiple comparisons are being performed (e.g., Duncan, 1955; O'Brien, 1983; Rothman, 1990).

PROBLEMS

12.1 In Chapter 5 the *VLSSperCapita.csv* data set was used to explore whether economic differences existed among seven nonoverlapping, distinct regions of the Socialist Republic of Vietnam. Both conditional plots and descriptive statistics were examined to begin understanding these differences.

The seven regions include two heavily populated areas known as the Red River Delta and Mekong Delta, respectively. The other regions are: Northern Uplands, North Central Coast, Central Coast, Central Highlands, and the Southeast. In the midst of the country's relative prosperity, the question of economic parity across regions, as measured by annual household per capita expenditures, remained largely unanswered.

    a) Refer back to Chapter 5. Based on the descriptive statistics and plots generate three hypotheses for two simple contrasts and one complex contrast, based on your preliminary examination. Test these three hypotheses using one of the approaches presented in Chapter 11 or 12. Be sure that you write a justification for the approach you chose.

    b) Compute the standardized mean difference as an effect size measure for both pairwise contrasts. Compute eta-squared as a measure of effect for the one complex contrast. Check the distributions of the variables involved with the contrasts. If the distribution of expenditures for regions in the pairwise contrasts is nonnormal, then execute the effect size computation using a robust estimator. Is there a robust estimator for eta-squared? Try to find an answer to this question by looking on the Internet.

References

Agresti, A. (2002). *Categorical data analysis* (2nd ed.). New York: Wiley.

Algina, J., Keselman, H. J., & Penfield, R. D. (2005). An alternative to Cohen's standardized mean difference effect size: A robust parameter and confidence interval in the two independent groups case. *Psychological Methods, 10*(3), 317–328.

American Educational Research Association. (2006). Standards for reporting on empirical social science research in AERA publications. *Educational Researcher, 35*(6), 33–40.

American Psychological Association. (2009). *Publication manual of the American Psychological Association* (6th ed.). Washington, DC: Author.

American Psychological Association. (2010). *Displaying your findings: A practical guide for creating figures, posters, and presentations* (6th ed.). Washington, DC: Author.

Andrews, D. F., Bickel, P. J., Hample, F. R., Huber, P. J., Rogers, W. H., & Tukey, J. W. (1972). *Robust estimates of location: Survey and advances*. Princeton, NJ: Princeton University Press.

Badley, G. (2002). A really useful link between teaching and research. *Teaching in Higher Education, 7*, 443–455.

Balanda, K. P., & MacGillivray, H. L. (1988). Kurtosis: A critical review. *The American Statistician, 42*(2), 111–119.

Becker, R. A. (1994). A brief history of S. In P. Dirschedl & R. Ostermann (Eds.), *Computational statistics* (pp. 81–110). Heidleberg: Physica Verlag.

Benjamini, Y., & Hochberg, Y. (1995). Controlling the false discovery rate: A practical and powerful approach to multiple testing. *Journal of the Royal Statistical Society. Series B (Methodological)*, *57*(1), 289–300.

Benjamini, Y., & Yekutieli, D. (2001). The control of the false discovery rate in multiple testing under dependency. *Annals of Statistics*, *29*, 1165–1188.

Beran, R. (1987). Prepivoting to reduce level error of confidence sets. *Biometrika*, *74*, 457–468.

Berger, J. O. (2003). Could Fisher, Jeffreys and Neyman have agreed on testing? *Statistical Science*, *18*, 1–32.

Blackett, P. M. S. (1968). Address of the president professor P. M. S. Blackett, O.M., C.H., at the anniversary meeting, 30 November 1967. *Proceedings of the Royal Society of London. Series B, Biological Sciences*, *169*(1016), v–xviii.

Boos, D. D. (2003). Introduction to the bootstrap world. *Statistical Science*, *18*(2), 168–174.

Bowman, A., & Azzalini, A. (1997). *Applied smoothing techniques for data analysis: The kernel approach with S-Plus illustrations*. London: Oxford University Press.

Box, G. E. P. (1979). Robustness in the strategy of scientific model building. In R. Launer & G. Wilkinson (Eds.), *Robustness in statistics* (pp. 201–236). New York: Academic Press.

Box, G. E. P., & Draper, N. R. (1987). *Empirical model-building and response surfaces*. New York: Wiley.

Bretz, F., Hothorn, T., & Westfall, P. (2011). *Multiple comparisons using R*. Boca Raton, FL: CRC Press.

Brewer, C. A. (1999). Color use guidelines for data representation. In *Proceedings of the Section on Statistical Graphics, American Statistical Association* (pp. 55–60). Alexandria, VA: American Statistical Association.

Brewer, C. A., Hatchard, G., & Harrower, M. A. (2003). Colorbrewer in print. *Cartography and Geographic Information Science*, *30*(1), 5–32.

Campbell, D. T., & Stanley, J. C. (1963). *Experimental and quasi-experimental designs for research*. Chicago: Rand-McNally.

Canty, A. J. (2002). Resampling methods in R: The boot package. *R News*, *2/3*, 2–7.

Carpenter, J., & Bithell, J. (2000). Bootstrap confidence intervals: When, which, what? a practical guide for medical statisticians. *Statistics in Medicine*, *19*(9), 1141–1164.

Chatterjee, S. K. (2003). *Statistical thought: A perspective and history*. Oxford: Oxford University Press.

Cleveland, W. S. (1984). Graphical methods for data presentation: Full scale breaks, dot charts, and multibased logging. *The American Statistician*, *38*(4), 270–280.

Cleveland, W. S. (1993). *Visualizing data*. Summitt, NJ: Hobart Press.

Cleveland, W. S., & McGill, R. (1983). A color-caused optical illusion on a statistical graph. *The American Statistician*, *37*(2), 101–105.

Cochran, W. G. (1965). The planning of observational studies in human populations. *Journal of the Royal Statistical Society (Series A)*, *128*, 134–155.

Cochran, W. G., & Rubin, D. B. (1973). Controlling bias in observational studies: A review. *Sankhyā*, *35*, 417–446.

Cohen, J. (1962). The statistical power of abnormal-social psychological research. *Journal of Abnormal and Social Psychology*, *65*(3), 145–153.

Cohen, J. (1965). *Statistical power analysis for the behavioral sciences*. New York: Academic Press.

Cohen, J. (1990). Things I have learned (so far). *American Psychologist*, *45*, 1304–1312.

Cohen, J. (1994). The Earth is round ($p < .05$). *American Psychologist*, *49*, 997–1003.

Cook, R. D., & Weisberg, S. (1999). *Applied regression including computing and graphics*. New York: Wiley.

Cooper, M. M. (2005). An introduction to small-group learning. In N. J. Pienta, M. M. Cooper, & T. J. Greenbowe (Eds.), *Chemists' guide to effective teaching* (pp. 117–128). Upper Saddle River, NJ: Pearson Prentice Hall.

Cournot, A. A. (1984). *Cournot's oeuvres complètes* (Vol. 1; B. Bru, Ed.). Paris: Vrin. (Reprinted from *Exposition de la théorie des chances et des probabilités*, 1843, Paris: Hachette.)

Crawley, M. J. (2007). *The R book*. Hoboken, NJ: Wiley.

Cummins, K. (2009). Tips on writing results for a scientific paper. *AmStat News, September*, 39–41.

D'Agostino, R., Jr. (1998). Propensity score methods for bias reduction in the comparison of a treatment to a non-randomized control group. *Statistics in Medicine*, *17*(19), 2265–2281.

Darlington, R. B. (1970). Is kurtosis really 'peakedness?'. *The American Statistician*, *24*(2), 19–22.

Dasu, T., & Johnson, T. (2003). *Exploratory data mining and data cleaning*. Hoboken, NJ: Wiley.

David, H. A. (1995). First(?) occurrence of common terms in mathematical statistics. *The American Statistician*, *49*(2), 121–133.

Davison, A., & Hinkley, D. V. (1997). *Bootstrap methods and their application*. New York: Cambridge University Press.

Dayton, C. M. (1998). Information criteria for the paired-comparisons problem. *The American Statistician*, *52*, 144–151.

Dayton, C. M. (2003). Information criteria for pairwise comparisons. *Psychological Methods*, *3*, 61–71.

Dayton, C. M., & Pan, X. (2005). Best subsets using information criteria. *Journal of Modern Applied Statistical Methods*, *4*, 621–626.

Deheuvels, P. (1977). Estimation nonparamètrique de la densitè par histogrammes gènèralisès. *Rivista di Statistica Applicata*, *25*, 5–42.

DiCicco, T. J. (1984). On parameter transformations and interval estimation. *Biometrika*, *71*, 477–485.

DiCicco, T. J., & Efron, B. (1992). More accurate confidence intervals in exponential families. *Biometrika*, *79*, 231–245.

DiCicco, T. J., & Efron, B. (1996). Bootstrap confidence intervals. *Statistical Science*, *11*(3), 189–212.

DiCicco, T. J., & Tibshirani, R. (1987). Bootstrap confidence intervals and bootstrap approximations. *Journal of the American Statistical Association*, *82*(397), 163–170.

Dixon, W. J., & Yuen, K. K. (1999). Trimming and winsorization: A review. *The American Statistician*, *53*(3), 267–269.

Duncan, D. (1955). Multiple range and multiple F tests. *Biometrics*, *11*, 1–42.

Dunn, O. J. (1961). Multiple comparisons among means. *Journal of the American Statistical Association*, *56*, 54–64.

Dwass, M. (1957). Modified randomization tests for nonparametric hypotheses. *The Annals of Mathematical Statistics*, *28*, 181–187.

Dyson, F. J., & Cantab, B. A. (1943). A note on kurtosis. *Journal of the Royal Statistical Society*, *106*(4), 360–361.

Eckhardt, R. (1987). Stan Ulam, John von Neuman, and the Monte Carlo method. *Los Alamos Science, Special Issue*, 131–143.

Edgington, E. S. (1966). Statistical inference and nonrandom samples. *Psychological Bulletin*, *66*(6), 485–487.

Edgington, E. S., & Onghena, P. (2007). *Randomization tests* (4th ed.). Boca Raton, FL: Chapman & Hall/CRC.

Efron, B. (1979). Bootstrap methods: Another look at the jackknife. *Annals of Statistics*, *7*, 1–26.

Efron, B. (1981a). Censored data and the bootstrap. *Journal of the American Statistical Association*, *76*, 312–319.

Efron, B. (1981b). Nonparametric standard errors and confidence intervals. *Canadian Journal of Statistics*, *9*, 139–172.

Efron, B. (1982). *The jackknife, the bootstrap, and other resampling plans*. Philadelphia, PA: Society for Industrial and Applied Mathematics.

Efron, B. (1987). Better bootstrap confidence intervals. *Journal of the American Statistical Association*, *82*, 171–200.

Efron, B. (1994). Missing data, imputation, and the bootstrap. *Journal of the American Statistical Association*, *89*(426), 463–475.

Efron, B., & Gous, A. (2001). Scales of evidence for model selection: Fisher versus Jeffreys. In P. Lahiri (Ed.), *Model selection* (pp. 208–246). Beachwood, OH: Institute of Mathematical Statistics.

Efron, B., & Tibshirani, R. J. (1993). *An introduction to the bootstrap*. New York: Chapman & Hall.

Epanechnikov, V. A. (1969). Nonparametric estimation of a multivariate probability density. *Theory of Probability and Its Applications*, *14*, 153–158.

Ernst, M. (2004). Permutation methods: A basis for exact inference. *Statistical Science*, *19*(4), 676–685.

Finucan, H. M. (1964). A note on kurtosis. *Journal of the Royal Statistical Society. Series B (Methodological)*, *26*(1), 111–112.

Fiori, A. M., & Zenga, M. (2009). Karl Pearson and the origin of kurtosis. *International Statistical Review*, *77*(1), 40–50.

Fisher, R. A. (1925). *Statistical methods for research workers*. Edinburgh: Oliver and Boyd.

Fisher, R. A. (1926). The arrangement of field experiments. *Journal of the Ministry of Agriculture*, *33*, 503–513.

Fisher, R. A. (1935). *The design of experiments*. Edinburgh: Oliver and Boyd.

Fisher, R. A. (1966). *The design of experiments* (8th ed.). Edinburgh: Oliver and Boyd.

Fox, J. (2005). The R commander: A basic-statistics graphical user interface to R. *Journal of Statistical Software*, *14*(9), 1–42.

Freedman, D., & Diaconis, P. (1981). On the histogram as a density estimator: L_2 theory. *Zeitschrift für Wahrscheinlichkeitstheorie und verwandte Gebiete*, *57*, 453–476.

Games, P. A. (1971). Multiple comparisons of means. *American Educational Research Journal*, *8*(53), 531–565.

Gardner, C. D., Kiazand, A., Alhassan, S., Kim, S., Stafford, R. S., Balise, R. R., et al. (2007). Comparison of the Atkins, Zone, Ornish, and LEARN diets for change in weight and related risk factors among overweight premenopausal women. the A TO Z weight loss study: A randomized trial. *Journal of the American Medical Association*, *297*(9), 969–977.

Gelman, A., Carlin, J. B., Stern, H. S., & Rubin, D. B. (2003). *Bayesian data analysis* (2nd ed.). London: CRC Press.

Gelman, A., Hill, J., & Yajima, M. (in press). Why we (usually) don't have to worry about multiple comparisons (with discussion). *Journal of Research on Educational Effectiveness*.

General Statistical Office. (2001). *Statistical yearbook*. Hanoi, Vietnam: Statistical Publishing House.

Gentleman, R. (2004). Some perspectives on statistical computing. *The Canadian Journal of Statistics*, *32*(3), 209–226.

Glass, G. V. (1977). Integrating findings: The meta-analysis of research. *Review of Research in Education*, *5*, 351–379.

Good, P. I. (2005a). *Introduction to statistics through resampling methods and R/S-Plus illustrations*. Hoboken, NJ: Wiley.

Good, P. I. (2005b). *Permutation, parametric, and bootstrap tests of hypotheses* (3rd ed.). New York: Springer.

Gottfredson, D. C., Cross, A., Wilson, D., Rorie, M., & Connell, N. (2010). An experimental evaluation of the All Stars prevention curriculum in a community after school setting. *Prevention Science*, *11*(2), 351–379.

Hall, P. (1986). On the bootstrap and confidence intervals. *The Annals of Statistics*, *14*, 1431–1452.

Hall, P. (1988). Theoretical comparison of bootstrap confidence intervals (with discussion). *The Annals of Statistics*, *16*, 927–985.

Hall, P. (2003). A short prehistory of the bootstrap. *Statistical Science*, *18*(2), 156–167.

Hall, P., & Martin, M. A. (1988). On bootstrap resampling and iteration. *Biometrika*, *75*, 661–671.

Hampel, F. R., Ronchetti, E. M., Rousseeuw, P. J., & Stahel, W. A. (1986). *The approach based on influence functions*. New York: Wiley.

Hancock, G. R., & Klockars, A. J. (1996). The quest for alpha: Developments in multiple comparison procedures in the quarter century since games (1971). *Review of Educational Research*, *66*(3), 269–306.

Harrower, M. A., & Brewer, C. A. (2003). Colorbrewer.org: An online tool for selecting color schemes for maps. *The Cartographic Journal*, *40*, 27–37.

Haughton, D., Haughton, J., & Phong, N. (2001). *Living standards during an economic boom: The case of vietnam*. Hanoi, Vietnam: Statistical Publishing House and United Nations Development Programme.

Henry, G. T., & MacMillan, J. H. (1993). Performance data: Three comparison methods. *Evaluation Review*, *17*, 643–652.

Hinkle, D. E., Wiersma, W., & Jurs, S. G. (2003). *Applied statistics: For the behavioral sciences* (5th ed.). New York: Houghton Mifflin.

Hinkley, D., & Wei, B. C. (1984). Improvement of jackknife confidence limit methods. *Biometrika*, *71*, 331–339.

Hochberg, Y. (1988). A sharper Bonferroni procedure for multiple tests of significance. *Biometrika*, *75*, 800–803.

Hochberg, Y., & Tamhane, A. C. (1987). *Multiple comparison procedures*. New York: Wiley.

Holm, S. (1979). A simple sequentially rejective multiple test procedure. *Scandinavian Journal of Statistics*, *6*, 65–70.

Hommel, G. (1988). A stagewise rejective multiple test procedure based on a modified bonferroni test. *Biometrika*, *75*, 385–386.

Hornik, K. (2010). *The R FAQ*. http://CRAN.R-project.org/doc/FAQ/R-FAQ.html. (ISBN 3-900051-08-9.)

Hothorn, T., Hornik, K., van de Wiel, M. A., & Zeileis, A. (2008). Implementing a class of permutation tests: The coin package. *Journal of Statistical Software*, *28*(8), 1–23.

Hsu, J. (1996). *Multiple comparisons: Theory and methods*. New York: Chapman & Hall.

Hubbard, R., & Bayarri, M. J. (2003). Confusion over measures of evidence (p's) versus errors (α's) in classical statistical testing (with discussion). *The American Statistician*, *57*, 171–182.

Huber, P. J. (1981). *Robust statistics*. New York: Wiley.

Huberty, C. J. (2002). A history of effect size indices. *Educational and Psychological Measurement*, *62*, 227–240.

Ihaka, R. (2003). Colour for presentation graphics. In K. Hornik, F. Leisch, & A. Zeileis (Eds.), *Proceedings of the 3rd international workshop on distributed statistical computing (DSC 2003)*. Vienna, Austria. http://www.ci.tuwien.ac.at/Conferences/DSC-2003/Proceedings/.

Institute of Education Sciences. (2008). *What works clearinghouse: Procedures and standards handbook (version 2.0)*. Washington, DC: United States Department of Education.

Joanes, D. N., & Gill, C. A. (1998). Comparing measures of sample skewness and kurtosis. *The Statistician, 47*(Part 1), 183–189.

Jones, M. C., Marron, J. S., & Sheather, S. J. (1996). A brief survey of bandwidth selection for density estimation. *Journal of the American Statistical Association, 91*(433), 401–407.

Kahn, H. (1954). Use of different Monte Carlo sampling techniques. In H. A. Meyer (Ed.), *Symposium on Monte Carlo methods* (pp. 146–155). New York: Wiley.

Kaiser, P. K., & Boynton, R. M. (1996). *Human color vision* (2nd ed.). Washington, DC: Optical Society of America.

Kempthorne, O. (1955). The randomization theory of experimental inference. *Journal of the American Statistical Association, 50*(271), 946–967.

Kempthorne, O. (1977). Why randomize? *Journal of Statistical Planning and Inference, 1*(1), 1–25.

Kempthorne, O. (1979). Sampling inference, experimental inference and observation inference. *Sankhya: The Indian Journal of Statistics, Series B, 40*(3/4), 115–145.

Kempthorne, O. (1986). Randomization—II. In S. Kotz & N. L. Johnson (Eds.), *Encyclopedia of statistical sciences* (Vol. 7, pp. 519–524). New York: Wiley.

Kempthorne, O., & Doerfler, T. E. (1969). The behavior of some significance tests under experimental randomization. *Biometrika, 56*(2), 231–248.

Keppel, G., & Wickens, T. D. (2007). *Design and analysis: A researcher's handbook* (5th ed.). Englewood Cliffs, NJ: Prentice Hall.

Kirk, R. E. (1995). *Experimental design: Procedures for the behavioral sciences* (3rd ed.). Pacific Grove, CA: Brooks/Cole.

Kirk, R. E. (1996). Practical significance: A concept whose time has come. *Educational and Psychological Measurement, 56,* 746–759.

Kirk, R. E. (2001). Promoting good statistical practices: Some suggestions. *Educational and Psychological Measurement, 61*(2), 213–218.

Koyama, T. (n.d.). *Beware of dynamite.* http://biostat.mc.vanderbilt.edu/wiki/pub/Main/TatsukiKoyama/Poster3.pdf.

Kruskal, W. (1988). Miracles and statistics: The casual assumption of independence. *Journal of the American Statistical Association, 83*(404), 929–940.

Kupfersmid, J. (1988). Improving what is published: A model in search of an editor. *American Psychologist, 43,* 635–642.

Leaf, M. (2002). A tale of two villages: Globalization and peri-urban change in china and vietnam. *Cities, 19*(1), 23–31.

Lee, S. M. S., & Young, G. A. (1995). Asymptotic iterated bootstrap confidence intervals. *The Annals of Statistics, 23,* 1301–1330.

Lee, V. S. (Ed.). (2004). *Teaching and learning through inquiry: A guidebook for institutions and instructors.* Sterling, VA: Stylus.

Lehmann, E. L. (1993). The Fisher, Neyman–Pearson theories of testing hypotheses: One theory or two? *Journal of the American Statistical Association, 55,* 62–71.

Leinhardt, G., & Leinhardt, S. (1980). Chapter 3: Exploratory data analysis: New tools for the analysis of empirical data. *Review of Research in Education,*

18(185), 85–157.

Liao, T. F. (2002). *Statistical group comparison*. New York: Wiley.

Light, R. J., Singer, J. D., & Willett, J. B. (1990). *By design: Planning research on higher education*. Cambridge, MA: Harvard University Press.

Lindley, D. V. (2000). The philosophy of statistics. *The Statistician, 49*(Part 3), 293–337.

Little, R. J. A. (1989). Testing the equality of two independent binomial proportions. *The American Statistician, 43*, 283–288.

Lochman, J. E., Lampron, L. B., Gemmer, T. C., Harris, S. E., & Wyckoff, G. M. (1989). Teacher consultation and cognitive-behavioral interventions with aggressive boys. *Psychology in the Schools, 26*(2), 179–188.

Loh, W.-Y. (1987). Calibrating confidence coefficients. *Journal of the American Statistical Association, 82*, 155–162.

Lohr, S. (2010). *Sampling: Design and analysis* (2nd ed.). Boston, MA: Brooks/Cole.

Lomax, R. G. (2007). *Statistical concepts: A second course* (3rd ed.). Mahwah, NJ: Lawrence Erlbaum Associates.

Long, J. D., Feng, D., & Cliff, N. (2003). Ordinal analysis of behavioral data. In J. Schinka & W. F. Velicer (Eds.), *Research methods in psychology. Volume 2 of handbook of psychology* (pp. 635–662). New York: Wiley.

Lord, F. M. (1953). On the statistical treatment of football numbers. *American Psychologist, 8*, 750–751.

Lumley, T. (2006). Color coding and color blindness in statistical graphics. *ASA Statistical Computing & Graphics Newsletter, 17*(2), 4–7.

Marriot, M. (2007, February 8). If Leonardo had made toys. *New York Times*.

Moore, D. S. (1990). Uncertainty. In L. A. Steen (Ed.), *On the shoulders of giants: New approaches to numeracy* (pp. 95–137). Washington, DC: National Academy Press.

Moors, J. J. A. (1986). The meaning of kurtosis: Darlington reexamined. *The American Statistician, 40*(4), 283–284.

Mosteller, F., & Tukey, J. W. (1977). *Data analysis and regression*. Reading, MA: Addison-Wesley.

Neyman, J. (1950). *First course in probability and statistics*. New York: Holt.

Neyman, J., & Pearson, E. (1928a). On the use and interpretation of certain test criteria for purposes of statistical inference. part i. *Biometrika, 20A*, 175–240.

Neyman, J., & Pearson, E. (1928b). On the use and interpretation of certain test criteria for purposes of statistical inference. Part II. *Biometrika, 20A*, 263–294.

Neyman, J., & Pearson, E. (1933). On the problem of the most efficient tests of statistical hypotheses. *Philosophical Transactions of the Royal Statistical Society of London, Ser. A, 231*, 289–337.

Oakes, J. M., & Johnson, P. J. (2006). Propensity score matching for social epidemiology. In J. M. Oakes & J. S. Kaufman (Eds.), *Methods in social epidemiology* (pp. 364–386). New York: Jossey-Bass.

O'Brien, P. C. (1983). The appropriateness of analysis of variance and multiple comparison procedures. *Biometrics, 39*, 784–794.

Oehlert, G. (2000). *A first course in design and analysis of experiments*. New York: W. H. Freeman.

Pearson, K. (1894). Contributions to the mathematical theory of evolution. *Philosophical Transactions of the Royal Society A*, *185*, 71–114.

Pearson, K. (1895). Contributions to the mathematical theory of evolution. II. Skew variation in homogeneous material. *Philosophical Transactions of the Royal Society A*, *186*, 343–414.

Pearson, K. (1901). On the correlation of characters not quantitatively measurable. *Philosophical Transactions of the Royal Society of London*, *195*, 1–47.

Pedhazur, E. J. (1997). *Multiple regression in behavioral research* (3rd ed.). Fort Worth, TX: Harcourt Brace.

Pitman, E. J. G. (1937). Significance tests which may be applied to samples from any populations. *Supplement to the Journal of the Royal Statistical Society*, *4*(1), 119–130.

Portnoy, S., & Koenker, R. (1989). Adaptive L estimation of linear models. *Annals of Statistics*, *17*, 362–381.

Prince, M. J., & Felder, R. M. (2006). Inductive teaching and learning methods: Definitions, comparisons, and research bases. *Journal of Engineering Education*, *95*(2), 123–138.

Quenouille, M. H. (1949). Approximate tests of correlation in time series. *Journal of the Royal Statistical Society, Series B*, *11*, 18–84.

Quenouille, M. H. (1956). Notes on bias in estimation. *Biometrika*, *43*, 353–360.

R Development Core Team. (2009). *R data import/export (version 2.10.1)*. Vienna, Austria: R Foundation for Statistical Computing.

Raspe, R. E. (1948). *Singular travels, campaigns and adventures of Baron Munchausen* (J. Carswell, Ed.). London: Cresset Press. (Original work published 1786.)

Rice, W. R. (1988). A new probability model for determining exact p values for 2×2 contingency tables (with discussion). *Biometrics*, *44*, 1–22.

Rissanen, J., Speed, T., & Yu, B. (1992). Density estimation by stochastic complexity. *IEEE Trans. on Information Theory*, *38*, 315–323.

Rosenbaum, P. R. (1995). *Observational studies*. New York: Springer.

Rosenbaum, P. R., & Rubin, D. B. (1983). The central role of the propensity score in observational studies for causal effects. *Biometrika*, *70*(1), 41–55.

Rosenberger, J. L., & Gasko, M. (1983). Comparing location estimators: Trimmed means, medians, and trimeans. In D. C. Hoaglin, F. Mosteller, & J. W. Tukey (Eds.), *Understanding robust and exploratory data analysis* (pp. 297–328). New York: Wiley.

Rosenthal, R. (1991). *Meta-analytic procedures for social research*. Thousand Oaks, CA: Sage.

Rosenthal, R., & Rosnow, R. L. (1985). *Contrast analysis: Focused comparisons in the analysis of variance*. Cambridge: Cambridge University Press.

Rosenthal, R., Rosnow, R. L., & Rubin, D. B. (2000). *Contrasts and effect sizes in behavioral research: A correlational approach*. Cambridge: Cambridge University Press.

Rosenthal, R., & Rubin, D. B. (1983). Ensemble-adjusted p values. *Psychological Bulletin, 94*(3), 540–541.

Rosnow, R. L., & Rosenthal, R. (1989). Statistical procedures and the justification of knowledge in psychological science. *Amercian Psychologist, 44*, 1276–1284.

Rosnow, R. L., & Rosenthal, R. (1996). Computing contrasts, effect sizes, and counternulls on other people's published data: General procedures for research consumers. *Psychological Methods, 1*, 331–340.

Rothman, K. (1990). No adjustments are needed for multiple comparisons. *Epidemiology, 1*, 43–46.

Rubin, D. (1981). The Bayesian bootstrap. *The Annals of Statistics, 9*, 130–134.

Sain, S. R. (1994). *Adaptive kernel density estimation*. Unpublished doctoral dissertation, Rice University.

Sarkar, D. (2008). *Lattice: Multivariate data visualization with R*. New York: Springer.

Saville, D. J. (1990). Multiple comparison procedures: The practical solution. *The American Statistician, 44*, 174–180.

Scheffé, H. (1953). A method for judging all contrasts in the analysis of variance. *Biometrika, 40*, 87–104.

Scott, D. W. (1979). On optimal and data-based histograms. *Biometrika, 66*, 605–610.

Senn, S. (2008). A century of t-tests. *Significance, 5*(1), 37–39.

Serlin, R. A., & Lapsley, D. K. (1993). Rational appraisal of psychological research and the good-enough principle. In G. Keren & C. Lewis (Eds.), *A handbook for data analysis in the behavioral sciences: Methodological issues* (pp. 199–228). Hillsdale, NJ: Erlbaum.

Shadish, W. R., Cook, T. D., & Campbell, D. T. (2002). *Experimental and quasi-experimental designs for generalized causal inference*. Boston, MA: Houghton Mifflin Company.

Shaffer, J. P. (1995). Multiple hypothesis testing. *Annual Review of Psychology, 46*, 561–584.

Shaver, J. (1985). Chance and nonsense: A conversation about interpreting tests of statistical significance, part 2. *Phi Delta Kappan, 67*(2), 138–141.

Sheather, S. J. (2004). Density estimation. *Statistical Science, 19*(4), 588–597.

Sheather, S. J. (2009). *A modern approach to regression with R*. New York: Springer.

Silverman, B. W. (1986). *Density estimation for statistics and data analysis*. London: Chapman and Hall.

Simon, J. L. (1997). *Resampling: The new statistics* (2nd ed.). Boston: Wadsworth.

Simonoff, J. S. (1996). *Smoothing methods in statistics*. New York: Springer.

Sohn, D. (2000). Significance testing and the science. *American Psychologist, 55*, 964–965.

Spector, P. (2008). *Data manipulation with R*. New York: Springer.

Stamps, K., & Bohon, S. A. (2006). Educational attainment in new and established Latino metropolitan destinations. *Social Science Quarterly, 87*(5), 1225–1240.

Staudte, R. G., & Sheather, S. J. (1990). *Robust estimation and testing*. New York: Wiley-Interscience.

Stevens, S. S. (1946). On the theory of scales of measurement. *Science, 103*(2684), 677–680.

Stevens, S. S. (1951). Mathematics, measurement, and psychophysics. In S. S. Stevens (Ed.), *Handbook of experimental psychology* (pp. 1–49). New York: Wiley.

Student. (1908). On the probable error of a mean. *Biometrika, 6,* 1–24.

Student. (1927). Errors of routine analysis. *Biometrika, 19*(1/2), 151–164.

Sturges, H. (1926). The choice of a class-interval. *Journal of the American Statistical Association, 21,* 65–66.

Tankard, J. W., Jr. (1984). W. S. Gosset and the *t*-test. In *The statistical pioneers* (pp. 87–109). Cambridge, MA: Schenkman Books.

Terrell, G. R. (1990). The maximal smoothing principle in density estimation. *Journal of the American Statistical Association, 85*(410), 470–477.

Terrell, G. R., & Scott, D. W. (1992). Variable kernel density estimation. *The Annals of Statistics, 20,* 1236–1265.

Thisted, R. A., & Velleman, P. F. (1992). Computers and modern statistics. In D. C. Hoaglin & D. S. Moore (Eds.), *Perspectives on contemporary statistics, MAA notes no. 21* (pp. 41–53). Washington, DC: Mathematical Association of America.

Thompson, B. (1994). *The concept of statistical significance testing* (Report No. EDO-TM-94-1). Washington, DC: Office of Educational Research and Improvement.

Thompson, B. (1996). AERA editorial polices regarding statistical significance testing: Three suggested reforms. *Educational Researcher, 25*(2), 26–30.

Thompson, B. (1997). Editorial policies regarding statistical significance tests: Further comments. *Educational Researcher, 26*(5), 29–32.

Thompson, B. (2007). Effect sizes, confidence intervals, and confidence intervals for effect sizes. *Psychology in the Schools, 44*(5), 423–432.

Tufte, E. R. (1990). *Envisioning information.* Cheshire, CT: Graphics Press.

Tukey, J. W. (1953). *The problem of multiple comparisons.* Unpublished manuscript. Princeton University.

Tukey, J. W. (1958). Bias and confidence in not-quite large samples. *The Annals of Mathematical Statistics, 29,* 614.

Tukey, J. W. (1962). The future of data analysis. *The Annals of Mathematical Statistics, 33,* 1–67.

Tukey, J. W. (1977). *Exploratory data analysis.* Reading, MA: Addison-Wesley.

Turlach, B. (1991). *Bandwidth selection in kernel density estimation: A review.* http://ideas.repec.org/p/wop/humbse/9307.html.

Upton, G. J. G. (1982). A comparison of alternative tests for the 2×2 comparative trial. *Journal of the Royal Statistical Society, Ser. A, 148,* 86–105.

Venables, W. N., & Ripley, B. D. (2002). *Modern applied statistics with s* (4th ed.). New York: Springer.

Venables, W. N., Smith, D. M., & the R Development Core Team. (2009). *An introduction to R. notes on R: A programming environment for data analysis and graphics (version 2.10.1)*. Vienna, Austria: R Foundation for Statistical Computing.

Wainer, H. (1984). How to display data badly. *The American Statistician, 38*(2), 137–147.

Wand, M. P. (1997). Data-based choice of histogram bin width. *The American Statistician, 51*, 59–64.

Wand, M. P., & Jones, M. C. (1995). *Kernel smoothing*. London: Chapman and Hall.

Westfall, P. H. (1997). Multiple testing of general contrasts using logical constraints and correlations. *Journal of the American Statistical Association, 92*(437), 299–306.

Westfall, P. H., & Young, S. S. (1993). *Resampling-based multiple testing*. New York: Wiley.

Wilcox, R. R. (2001). *Fundamentals of modern statistical methods: Substantially improving power and accuracy*. New York: Springer.

Wilcox, R. R. (2004). Kernel density estimators: An approach to understanding how groups differ. *Understanding Statistics, 3*(4), 338–348.

Wilcox, R. R. (2005). *Introduction to robust estimation and hypothesis testing* (2nd ed.). San Diego, CA: Academic Press.

Wilkinson, G. N., & Rogers, C. E. (1973). Symbolic description of factorial models for analysis of variance. *Applied Statistics, 22*, 392–399.

Wilkinson, L. (2006). *The grammar of graphics*. New York: Springer.

Williams, V. S. L., Jones, L. V., & Tukey, J. W. (1999). Controlling error in multiple comparisons, with examples from state-to-state differences in educational achievement. *Journal of Educational and Behavioral Statistics, 24*(1), 42–69.

Winch, R. F., & Campbell, D. T. (1969). Proof? no evidence? yes. the significance of tests of significance. *The American Sociologist, 4*(May), 140–144.

WorldBank. (1999). *World development indicators 1999*. CD-ROM. Washington, DC: WorldBank.

Yates, F. (1984). Tests of significance for 2×2 contingency tables (with discussion). *Journal of the Royal Statistical Society, Ser. A, 147*, 426–463.

Zahn, I. (2006). Learning to Sweave in APA style. *The PracTeX Journal, 1*.

Zeileis, A., & Hornik, K. (2006, October). *Choosing color palettes for statistical graphics* (Report No. 41). Department of Statistics and Mathematics, Wirtschaftsuniversit at Wien, Research Report Series.

Zeileis, A., Hornik, K., & Murrell, P. (2007, November). *Escaping RGBland: Selecting colors for statistical graphics* (Report No. 61). Department of Statistics and Mathematics, Wirtschaftsuniversit at Wien, Research Report Series.

Zhang, H., & Montag, E. D. (2006). How well can people use different color attributes. *Color Research & Application, 31*(6), 445–457.

Printed and bound by CPI Group (UK) Ltd, Croydon, CR0 4YY

27/10/2024

14580252-0001